Simplicius

On Aristotle's Physics 4.1-5, 10-14

Simplicius

On Aristotle's Physics 4.1-5, 10-14

Translated by J. O. Urmson

Cornell University Press
Ithaca, New York

For information address Cornell University Press, 124 Roberts Place, Ithaca, New York 14850.

First published 1992 by Cornell University Press.

Library of Congress Cataloging-in-Publication Data

Simplicius, of Cilicia.
On Aristotle's Physics 4.1-5. 10-14 / Simplicius ; translated by J. O. Urmson.
p. cm. — (Ancient commentators on Aristotle)
Includes bibliographical references and index.
ISBN 0-8014-2817-3
1. Aristotle. Physics. 2. Place (Philosophy)—Early works to 1800. 3. Time—Early works to 1800. I. Urmson, J. O. II. Title. III. Series.
Q151.A8S53 1992 92-25477

Printed in Great Britain

Contents

Introduction

Richard Sorabji

This book is a companion to J.O. Urmson's translation of Simplicius' *Corollaries on Place and Time*. Simplicius separated off his own theories of place and time and their pre-history, and treated them in two corollaries distinct from the main commentary. The commentary itself is translated here; thanks to the separation, it is much more directly addressed to Aristotle.

Aristotle's account of place in the first five chapters is surprising, because it makes place two-dimensional. My place is the (two-dimensional) inner surface of my surroundings.[1] Aristotle also thinks that the upward motion of air and fire and the downward motion of the other two elements, earth and water, is partly explained by the natural places to which they tend. Place thus has power (*dunamis*) of its own.[2]

Both views were questioned by Aristotle's immediate successor Theophrastus, whose arguments Simplicius records in the *Corollaries*.[3] The first in particular, the idea of place as a two-dimensional surface, remained unattractive to most Greek thinkers, and was revived only in the Latin West of the Middle Ages.

Aristotle's account of time in the last five chapters is challenging in a different way, because he starts off with two fascinating puzzles about whether time exists. None of its parts exists, if the past exists no longer, the future does not yet exist, and the present, being a sizeless instant, is not a part. Moreover, there is no time at which the present instant can ever cease to exist. It cannot cease while it is present, it cannot last for a while and cease at a later instant, nor, as Aristotle rightly perceives, is there such a thing as the very next

[1] Aristotle *Physics* 4.4. At 4.1, 209a5, Aristotle's statement that place is three-dimensional is not endorsed, I believe, but is used to show that the conventional view raises a puzzle (how place differs from body). However, for a different view, see note 22 to the translation below.

[2] Aristotle *Physics* 4.1, 208b11.

[3] Simplicius *in Phys*. 604,5-11; 639,13-22; Richard Sorabji, *Matter, Space and Motion*, London and Ithaca N.Y., 1988, chs. 11-12.

instant, if time is a continuum.[4]

In giving a positive account of time, Aristotle argues that it must involve change, or its passage would be undetectable.[5] I take it that such a sceptical hypothesis would be incompatible with Aristotle's methodology of starting from received opinions.

Aristotle suggests that time is something we count when we count different stages, some before and some after, in a movement, particularly in the movement of the heavens.[6] Their rotation supplies the change we need. Further, time requires consciousness. For time, being by definition countable, requires the existence of beings to do the counting.[7]

Simplicius was writing over eight hundred years later, after the closure of the Athenian Neoplatonist school in AD 529, possibly in Harrân, which is in present-day Turkey.[8] His commentary is rich in reports of earlier discussions. He is able, for example, to check Aristotle's report of Zeno's puzzle, that every place will need a place *ad infinitum*. He does so both by noting inconsistencies in Aristotle's own wording and by drawing on the different report of Zeno handed down by Aristotle's pupil Eudemus of Rhodes.[9]

Some of the best criticisms of Aristotle on place are those of Theophrastus, which Simplicius reserves for the *Corollaries*, but many of the best comments on Aristotle's treatment of time are kept for the commentary.

Galen, the great medical scientist of the second century AD, attacked Aristotle's argument that time requires change. But as I understand the reports of Simplicius and Themistius,[10] he wrongly took Aristotle to be saying that, because we have to change (*kinoumenoi*: Themistius) in our minds when we think, we are obliged to think of time as changing (*kinoumenon*: Simplicius). He objects that in that case we would be obliged to think of everything as changing, even such unchanging things as the centre of the earth (the example in Themistius). You might as well say that if we have to think of an unextended point, for example, by using an extended mental image, we have to think of the point as extended, but in fact the representation need not be like the represented. It is easiest, I

[4] Aristotle *Physics* 4.10. I have argued (*Time, Creation and the Continuum*, London and Ithaca N.Y., 1983, 10-12) that Aristotle's solution to the second is given at *Metaphysics* 3.5, 1002a28-b11. I believe the first is left unsolved.

[5] Aristotle *Physics* 4.11, 218b21-219a1.

[6] Aristotle *Physics* 4.11, 219a1-b9.

[7] Aristotle *Physics* 4.14, 223a21-9.

[8] This is the suggestion of Michael Tardieu, 'Sâbiens coraniques et "Sâbiens" de Harrân', *Journal Asiatique*, 274, 1986, 1-44, fully reported by I. Hadot in Richard Sorabji, ed., *Aristotle Transformed: the Ancient Commentators and their Influence*, London and Ithaca N.Y., 1990, ch. 12.

[9] 563,8ff.

[10] Simplicius 708,27; Themistius *in Phys*. 144,24ff.

think, to take this last remark about unextended things as being part of Galen's objection. On the punctuation of Diels, the editor, it is part of Simplicius' reply to Galen, but in that case Simplicius will have misunderstood. The translator has kept to Diels's text, so that the reader may judge for himself or herself which is the right interpretation.

Simplicius records the striking refutation offered by Eudemus of Rhodes against those Pythagoreans who said that history would repeat itself exactly again and again.[11] This would not be a repetition, so I take Eudemus to reply, because the time would be the same, not different. He is relying on the idea that, since the countable stages in what is going on are supposed to be the same, any count would come out the same, so that, by Aristotle's definition of time, the time would be the same.

Sometimes Simplicius has a better solution to offer than his predecessors. He improves, for example, on Themistius' explanation of how the eternal rotation of the heavens can be considered as falling within time, when it is no shorter than time. Simplicius' idea seems to be that however large a finite number of rotations you take, there will always be time longer than that.[12]

Aristotle's argument for time involving consciousness came under repeated scrutiny. There is a parallel with his argument that place involves the possibility of receiving bodies, and hence that there is no place beyond the furthermost stars, because there is no possibility of matter flying out to be received.[13] Similarly, time involves the possibility of being counted, and hence there is no time in the absence of consciousness, because there is no possibility of any counting. Is it true that the absence of one possibility (that of counting or being received) implies the absence of the other (that of being counted or receiving)? I think not in the relevant sense, because the absence of any *opportunity* for the first does not imply the absence of a *capacity* for the second. But different views were taken in antiquity.

Boethus of Sidon in the first century BC is reported as rejecting Aristotle's view. The countable can exist without someone to do the counting, just as the perceptible can exist without a perceiver.[14] Alexander of Aphrodisias, however, who held an Aristotelian chair around AD 205, defended his master.[15] Simplicius agrees,[16] but

11 732,26-733,1.

12 741,22ff.

13 Aristotle *de Caelo* 1.9, 279a11-18. See *Time, Creation and the Continuum* 89-93; *Matter, Space and Motion*, 132-5.

14 766,17-19 (=Themistius *in Phys.* 163,5-7) and 759,18-20.

15 759,20-760,3; *de Tempore* (Théry) 95,11-12, translated by R.W. Sharples in *Phronesis* 27, 1982, 58-61, at 64.

16 760,33-761,5.

elsewhere he is conscious that Aristotle himself did not always take the same view about correlative possibilities. In the *Categories* Aristotle allows that there might be perceptibles and knowables without any perceivers or knowers, whereas Simplicius rejects any such asymmetry.[17]

While Simplicius reserved his own theories and their historical background for the *Corollaries*, this has not prevented the commentary translated here from being a major sourcebook for the interpretation of Aristotle.

Some of the footnotes to the translation are the responsibility of the general editor; the translator's footnotes are indicated by '(J.O.U.)'.

Acknowledgments

The present translations have been made possible by generous and imaginative funding from the following sources: the National Endowment for the Humanities, Division of Research Programs, an independent federal agency of the USA; the Leverhulme Trust; the British Academy; the Jowett Copyright Trustees; the Royal Society (UK); Centro Internazionale A. Beltrame di Storia dello Spazio e del Tempo (Padua); Mario Mignucci; Liverpool University. I further wish to thank Ian Crystal and Paul Opperman for their help in compiling the indexes and preparing the volume for press.

[17] Simplicius *in Cat.* 196,12.27-33.

Textual Emendations

526,2 Extending the quotation from Diels's 526,2 after *topôi* to 526,4 after *diaphoras*.

530,26 Replacing a full stop by a question mark.

530,29 Replacing a question mark by a full stop.

538,13 Reading e.g. *diaspasmon* for *parainesin*.

538,18 Reading *ei gar* with a for the *ou gar* of EF.

542,30-1 Correcting the end of the lemma from *heôs tou to de angeion ouden tou pragmatos estin* to *heôs tou allôn sômatôn homoiôs*.

543,11-12 Correcting the end of the lemma from *heôs tou to de angeion ouden tou pragmatos estin* to *alla khôristos ho topos hekastou esti*.

545,8 Reading *tou en hôi ektos ontos* with aF instead of *to en hôi ektos on*.

551,7 Reading *toutous* for *toutois*, which appears to be a misprint.

554,29-30 Omitting *kai hoti kat' allo adunaton* with aF.

557,21.22 Reading *en hautôi* for *en autôi* in lines 21 and 22.

563,8 Reading a question mark for a full stop.

568,13 Accepting Diels's conjecture *<dia ti> de* (sc. *dêlon*) *hoti*.

576,3 Assuming some such conjecture as Diels's *kai ho eis hon* for the impossible *ei hos* of E.

584,14 Conjectured closing of a quote that Diels fails to close, after *hudatos*.

587,13-14 Reading *periekhei kai topon lambanei* for *periekhei, topon lambanei*.

591,13-14 Reading e.g. *diaireta* between *mêpô* and *mêrê*.

592,22 Eliminating the comma between *pôs êkousen* and *hôs eirêtai*.

599,27 Reading *tote* for *touto* as Diels suggests in his apparatus.

699,24 Reading *epi* for *epei*.

706,34 Adding to lemma *mêden de diaphereto legein hêmin en tôi paronti kinêsin ê metabolên*,

	omitted by Simplicius but discussed under this lemma.
707,13	Reading *suntonous* for *sun tonous*, a misprint.
708,32	Deleting quotation marks after *noêsei nooumen*.
708,34	Inserting quotation marks after *phantastikôs energountes*.
717,13	Reading *esti* for *este*, with Spengel and Diels. Possibly read *esti dê pantakhê kai homoiôs arithmos hêstinos kinêseôs*, which gives the right sense.
717,32	Reading a question mark for a full stop.
721,1	Reading a question mark for a full stop.
721,4	Reading a question mark for a full stop.
723,27	Reading *topon* for *tropon*.
734,2	Reading a full stop for the comma after *pollê*.
738,34	Reading a question mark for a full stop.
744,15	Reading *ekeinou* for *ekeinês*, as suggested by Diels in his apparatus.
747,32	Deleting *to men oun houtô legetei tôn nun*.
752,17	Adding quotation marks before *houtô*.
768,28	Reading *ton prôton* for *tên prôtên*.
770,17	Adding *kai tôn hepta hippôn* after *kunôn*.

Simplicius

On Aristotle Physics 4.1-5, 10-14

Translation

Translator's Note

This is a translation of Simplicius' commentary on chapters 1-5 and 10-14 of Aristotle's *Physics* Book 4. The Greek text is that edited by Diels in the *Commentaria in Aristotelem Graeca*, volume 9, Berlin 1882, except for a few small emendations, a list of which appears on pp. 5-6 of this volume. The text discusses short passages of Aristotle, one after another, the passage being indicated in a lemma containing only the first few and the last few words. The numbers from 207 to 224 in these lemmata are page numbers in Aristotle's *Physics*. For the convenience of the reader this translation contains the whole of these passages, the parts omitted from the lemma being enclosed in square brackets. I have used my own translation, rather than borrowing one of those already available, first in order to ensure uniformity of vocabulary between text and commentary, and secondly because the text of Aristotle supplied in the lemma differs in some places certainly, in some probably, from the text of Ross's edition of the *Physics*, which I have otherwise followed. These differences, usually slight, are noted when they occur. Marginal numbers are the page and line numbers of Diels's Greek text.

I am very grateful to Professor Sorabji and his assistants, Paul Opperman and Ian Crystal, both for saving me from many mistakes of translation and for their great skill and care in preparing this translation for press.

The Commentary of Simplicius on Book 4.1-5, 10-14 of the *Physics* of Aristotle

PROOEMIUM

[The commentary begins with a brief introduction showing how Book 4 follows logically after Books 1-3. – Tr.]

Having embarked on the investigation of the first principles and the 519,3
causes of natural things and of what is universally present in the whole of nature, in his first book Aristotle discussed those principles which function as elements – matter, form and its corresponding privation. In the second book he discussed the efficient and the final cause and certain things which seem to some to be in themselves efficient causes but which are not as such efficient, such as chance and spontaneity, but only contingently so. Having shown that nature is an efficient cause, he defined it as the principle of change[1] for bodies, since change is present in all natural phenomena *qua* natural and holds natural bodies together and is included in the definition of nature. So it was reasonable that he should discuss change immediately at the beginning of Book 3. But since it is necessary that every natural body and every change should be either limited or unlimited, and since both body and change, being

[1] The translation of the Greek word *kinêsis* is of great difficulty for the translator who wishes to avoid different translations of such a key word in different contexts. The problem is that in non-technical Greek it means 'motion', and is often used in that sense by Aristotle and his commentators. But Plato proposed extending its use (*Theaet.* 181) and Aristotle used it of all change in all the categories except for substance, relation and action-and-passion (*Metaph.* 1068a8). Thus the word 'change' is too wide as a translation, and is unsuitable when motion is clearly meant. In any case, 'change' is the natural translation of *metabolê*. But the traditional translation by 'motion' throughout requires qualitative change, for example, to be called 'motion in respect of quality' which is not English. Hence, 'motion' and 'change' are both used as translations of *kinêsis*. *kinêsis kata topon* is translated as 'change of place'; *kinêsis* is translated as 'change' except when it clearly refers to motion exclusively, when it is translated as 'motion'. *metabolê* is translated always as 'change'. The Greek-English index will indicate where *kinêsis* (cognate verb *kinein*, *kineisthai*) has been translated one way or the other, and where *metabolê* (and its cognate verb *metaballein*) occurs. For the translation of other words of change and motion, see the index, s.v. *allassein*, *alloioun*, *alloiôsis*, *ameibein*, *exallassein*, *metabainein*, *metabasis*, *metastasis*, *methistanai*, *phora*, *pheresthai*, *metapheresthai*. (J.O.U.)

continuous, are divisible without limit, and, furthermore, since some natural scientists have said that the unlimited was the first principle of nature, he reasonably also discussed the unlimited. Having completed his account of the elements and the other causes, and the further account of wrongly assumed causes, as the next stage he investigates and teaches concerning the external concomitants of natural bodies as such, examples being place and time. For every natural body is in a place, and it changes its place and remains in a place, while simple bodies are specified by their motion towards their proper places.[2] Also, since every change is measured temporally, he had to provide a doctrine of time, if indeed time is, as we shall learn, the number of all change.

Reasonably, he discourses first about place. For place precedes not only time but also change and body itself. For a body is in a place, and
520,1 change happens to a body, and time is present in change. But since the pre-eminently natural entity is the natural body and the rest either receive it, or are its change or are the measure of its change, he reasonably first gave an account of the elements of natural body – matter and form – then of the efficient and final causes of body, and of change which is included in the definition of nature. And thus he put place before time among the external concomitants.

But, since the natural scientist has things in change as his subject-matter, just as the mathematician and the theologian have the unchanging, and yet place is something unchanging, as will be shown, how could place be within the scope of the natural scientist? Perhaps it is because it receives changing bodies, even if it itself is unchanging, and provides a boundary for a changing body, *qua* changing, even if it does not itself change. But since some have put the void among first principles, and have said that it is place lacking a body, Aristotle reasonably discusses the void also after his discussion of place. Then he adds the problems concerning time, and thus completes the book we have before us and hence the discussion of the features universally present in nature, which is concerned with the first principles of nature.

CHAPTER 1

208a27-9 Similarly it is requisite that the natural scientist should ascertain whether place and the unlimited exist or not, what kind of existence they have, and what they are.

[2] The simple bodies, earth, water, air and fire, essentially seek to be positioned in that order, earth at the bottom, fire at the top. (J.O.U.)

The problems about place are defined by their similarity to the enquiries about the unlimited.[3] This is reasonable; for just as when we hypothesised either the existence or the non-existence of the unlimited, absurdities seemed to follow, so it is necessary for those who investigate whether place exists to probe in both these directions, since the problem is a novelty. Also the first among the problems about place is to examine whether it exists or not, as in the case of the unlimited. We have learnt from the lectures *On Demonstration*[4] that after whether it exists we should enquire what it is. For this lays bare the nature and the essence and the definition of the subject matter. Next follows the problem of the thing's qualities, which makes clear what its characteristics are, to which is related the question of its manner of existence. For as we said with regard to the unlimited that it exists potentially, bit by bit and in becoming, we similarly enquire into the manner of the existence of place, whether it is self-subsistent, the view of those who say that place is an interval or void, or whether it is relational, and whether it contributes something or nothing to the existence of bodies. For these questions are involved in those of the characteristics and manner of existence of place.

But why did Aristotle put the question of manner of existence earlier? Perhaps it was because we often discover a thing's essential 521,1
nature from its characteristics, as will now become clear in the case of place.

> **208a29-32** For we all suppose that things that exist are somewhere [(for we suppose that the non-existent is nowhere; for where is the goat-stag or the sphinx?), and of changes the most common and most basic is that of place,] which we call locomotion.

Having at the beginning defined the problems about place, and their likeness to those about the unlimited, he goes on to show that a discussion of place is necessary for the natural scientist by two arguments, of which one seems to be a received opinion derived from common sense, the other demonstrative and based on actual facts. The one based on common sense employs the hypothetical syllogism[5] put forward by the ancients, which runs as follows: if that

[3] The unlimited was discussed in Book 3.

[4] Presumably the work now called the *Posterior Analytics*. (J.O.U.)

[5] Hypothetical syllogisms used premises different from Aristotle's standard 'all', 'no', and 'some' premises. The premises included 'if, then', 'either, or', 'not both'. For a history, see Katerina Ierodiakonou, 'The Stoic indemonstrables in the later tradition', a paper presented at the symposium on Stoic logic in Bamberg, Germany, September 1991, and forthcoming in a collection of the symposium papers (eds During and Ebert).

which is not is nowhere, that which is is somewhere; but, if being somewhere makes evident the occupation of a place, it is evident that place exists. 'But,' as Alexander says, 'this supposition is not common to everyone, but only to those who think that only bodies exist, and only things having a matter, and who do away with incorporeal nature.' But, as it appears, the terms are converted irregularly; for from 'if that which is not is nowhere' it follows according to the second form of hypothetical argument[6] that what is somewhere is, not that what is is somewhere, which is what is required by those who are arguing for place. So we must either transpose the text, saying that that which is nowhere does not exist, and thus go on to say that the existent is somewhere, or else, more persuasively in the opinion of those using the argument who believe that only bodies exist, one should say that the antecedent and the consequent[7] are logically equivalent. For if there is nothing other than bodies it is clear that that which is nowhere does not exist at all and what does not exist is nowhere, since bodies are somewhere. In the case of equivalent propositions the order is indifferent, whether from the antecedent or from the consequent. It seems that Aristotle is parodying Plato's words in the *Timaeus* where he says: 'gazing on this we dream and say that it is necessary for the whole of what is to be somewhere in some place and to occupy some region, and that what is not in the earth or in the heavens is nothing.'[8] For you see that this argument is both straight and validly converted. For because that which is nowhere does not exist, therefore everything that exists is somewhere. But Plato does not say this as accepting the argument, but because from the dream-like gaze at the material world, of which this holds, we predicate it of all that is.

Aristotle shows that the non-existent is nowhere by asking 'For where is the goat-stag or the sphinx?' For they are nowhere since they do not exist at all. And they do not exist at all since they are nowhere. It is clear that this consequence is valid when thus construed, but it is not true. For neither is everything that exists in
522,1 a place (for nothing that is most truly immaterial and separated is in a place), nor is that which is nowhere, i.e. not in a place, non-existent. For, as has been said, there are many things that are not in a place.

But the discussion has shown what it set out to show, that a discussion of place is necessary for the natural scientist, if, indeed,

[6] This is the second of the five indemonstrable arguments found ascribed to the Stoic Chrysippus by Diogenes Laertius, *Lives of Eminent Philosophers* 7.76-81. His example is: If it is day, it is light. But not: it is light. Therefore not: it is day.

[7] Antecedent and consequent correspond to the 'if' and 'then' parts of the conditional.

[8] Plato *Timaeus* 52B. (J.O.U.)

the earlier ones think that all that exists is in a place. Now, if we are not to denounce these early men pointlessly, it is likely that 'somewhere' does not indicate a relation in the sort of place now being investigated, but a determination of order, just as Plato recognised a signified and supercelestial[9] place, and as we say that intelligence is in the intelligible place.

Having thus shown from the ancient scientists that the discussion of place is necessary, in the second argument he demonstrates the same point on the basis of the actual facts as follows: if the change common to all natural bodies is change of place, it is clear that place is a natural phenomenon; so the discussion of place is necessary for the natural scientist. But the antecedent, so the consequent.[10] He called this change the common one either (a) because it is commonly recognised by everyone. For we all perceive change of place more readily than other changes. That is why some have even denied the reality of passing away, like the school of Anaxagoras, or of qualitative change, like those who say that 'colour is conventional'.[11] But nobody speaking seriously has denied the existence of change of place (for we may disregard Zeno's paradox). Or (b) because this change alone is common to all natural bodies. For the heavenly bodies, that are exempt from all other change, do change in this way. This is why he calls it the most basic case of change. For the most basic case is that which is applicable to the most fundamental and originative things of all. Moreover, Aristotle will show later on that locomotion is the prime change temporally, naturally and in causal order, and such change is rightly called basic.

In some transcriptions 'prime' is written with 'common', in some 'prime' instead of 'common', and that is what Eudemus writes.[12]

> **208a32-b1** There are many problems about what place is. [For it does not seem to be the same thing when examined in the light of all the data. Nor do we have statements of and answers to the problems] from earlier thinkers.

Having shown that a discussion of place is necessary for the natural scientist, he goes on to show, as in other matters, that the problem is difficult and consequently in need of continuous application and pardon if an accurate answer does not immediately appear to the first persons to examine it systematically. Therefore the demonstration of the difficulty is useful for learner and teacher. He exhibits the difficulty of it first from the need to collect the

[9] The place of Forms in Plato's *Phaedrus*.
[10] A Stoic mode of presenting arguments.
[11] Democritus Fr. 9. (J.O.U.)
[12] Aristotle *Physics* 208b31. (J.O.U.)

523,1 definitions of investigated objects from their essential features, as he said at the beginning of *On the Soul*,[13] while what seem to be the essential features of place do not point to the same nature. For in that containment is a feature of place, the place of each object will seem to be its form;[14] for that is what immediately bounds and contains it. But in that it receives a certain form and seems to be its space it seems to be more like matter. This is why Plato speaks of matter when he seems to be talking about space and place. But in that the same place will receive a different body at different times it would seem to be something distinct from that which is in the place, and so other than its matter or form. In this way the essential features of place, leading as they do in different directions, make understanding of it troublesome.

But he informs us that a second cause of the difficulty of the subject is that nobody previously had raised the problem or offered solutions to it. This is both an indication and a cause of its being troublesome to understand. It is an indication in that people avoided the problem because of its difficulty, and it is a cause because the discussion becomes even more difficult for us when we have no collaborator in our enquiry nor find any sort of basis already laid down. For even if Hesiod apparently said

> Chaos came into existence first.[15]

Hesiod's contribution is mythical and not scientific. But this much, if anything, he showed well, that something must already be there in which things can come into being. Plato, in the *Timaeus*, when he appears to be discussing place, in fact spoke about matter.

Eudemus says that a further cause of the troublesomeness of the problem of place is that it is not easy to lay hold on it, because it altogether escapes us when the body in it is removed, and it is not possible to apprehend it in itself, but, if at all, in combination with something else, like the sounds of the so-called consonants. For with 'a' added the sound of 'b' and 'c' becomes clear. However, we should not shirk the problem because of its difficulty when it is so relevant and necessary for the natural scientist. For even if nothing more is

[13] As Diels remarks, it is not clear where he said this. (J.O.U.)

[14] Simplicius' own account of place in his *Corollary* incurs this difficulty that a thing's place is hard to distinguish from its form. His answer is that place gives order to forms (629,13-19; 630,24-30; 638,26-7, translated by Urmson). Aristotle never gave such a dynamic sense to containment and so avoids this problem. It was Iamblichus' reinterpretation of Aristotle in terms of Platonic Forms that foisted the problem on him. See Richard Sorabji, *Matter, Space and Motion* (London and Ithaca N.Y., 1988), ch. 12. For Aristotle's discussion of the relation of form to place with Simplicius' comments, see 535,13-537,4; 572,26-31.

[15] Hesiod *Theogony*, line 116. (J.O.U.)

achieved with what is hard to grasp, still to have raised the problem well is enough for the lover of understanding as Aristotle himself teaches elsewhere.

> **208b1-8** That place exists, [then, seems obvious because of successive displacement. For where there is now water, there, when it has gone as from a vessel, air is present instead and at another time some other body occupies that same place. This seems to be something different from everything that comes into it or replaces it. For where there is now air, there water was formerly, so that it is clear that the place and the space must be something different from both,] into which and from which they passed.

Having shown the necessity and the difficulty of the enquiry concerning place, it remains for him to move on to its specific 524,1
problems, of which the first is whether place exists. He sets out first the arguments that make the existence of place probable, and afterwards those that eliminate it. He expounds four arguments that affirm place. The first of these is that from successive displacement. For if, where formerly there was water, the water has run out and there is now air, just as happens in vessels, it is clear that what receives them is other than either of them. The second is from the natural locomotion of the elements. For each of these is specified by its locomotion, one upwards, another downwards; these specific differences are of place. The third is a proof from mathematical considerations according to Alexander, but I think he is mistaken. For the argument based on mathematics is a part of the second argument from the natural locomotion of the elements, as we shall learn. The third is in fact a dialectical argument from the void; for, if the void is a place lacking a body, those who say there is a void will certainly endorse the existence of place. As a fourth argument in addition to these he adds Hesiod's mythical poetry that shows not only that place exists but also that it has great power. It is clear that the first two arguments appeal to the necessity of the situation; the others are derived from the opinion of those who accept such hypotheses. But now that we have given a synoptic account of the arguments, we shall take up the thought of each and articulate it so far as we can.

The proof from successive displacement went somewhat as follows: a place will receive successively different objects while remaining the same. But what will receive successively different objects while remaining the same is something and a different thing from these objects, since it remains the same and they are different and at different times. So place is something and other than what it contains. That a place receives different things at different times while remaining the same he showed by successive displacement of bodies, saying 'where there is now air, formerly there was water'. From which it is clear that a space is something else into which and from which the bodies in it passed. But if it is different it clearly exists and is worth investigating. The comparison with the vessel added actuality to the proof of the existence and the otherness of that which receives its successive occupants. For the vessel is both something and also other than the things in it, which fact reinforces the major premiss.

> **208b8-25** Also the local motions of natural and simple bodies, [such as fire, earth and the like not merely show that place is something but also that it has a certain capacity. For each of them moves to its own place if not prevented, one up, the other down. But up and down and the rest of the six directions are parts and forms of place. Moreover up and down, to left and to right and the like are not merely relative to us. For to us they are not always the same, but depend on our position and how we turn, which is why the same thing can often be to right and to left, above and below, and in front and behind. But in nature each of these is distinct. For above is not anywhere you like but to where fire and light things move locally; similarly below is not anywhere you like but the place of heavy and earthy things. So they do not differ from each other only in position but also in their powers. Mathematical objects also illustrate the point; not being in a place they still have a position relative to us, to right and to left, but only relatively speaking, since] they do not have these properties naturally.

Also a second proof of the existence of place is based on the natural 525,1 motion of bodies. At the same time it demonstrates that place has some power in its own nature, so that for this reason also it is worth investigation. Aristotle shows that place is something and has a power as follows: if each natural and simple body moves locally to its proper place if not prevented (fire upwards, earth downwards and intermediate things to the middle, and what moves in a circle[16] moves round the middle point), and each is specified in this way, and if up and down and the other directions are parts and forms of place, then it is plain that place is something and has a power. Otherwise these differences would not specify simple bodies. But right and left, and in front and behind characterise celestial motion also, as Aristotle proves in *de Caelo*, and also living things in the world of

[16] The heavens.

becoming, in accordance with their composite nature.[17] Further, warmth and softness and similar attributes might exist without place, but it is impossible for there to be lightness and heaviness without place; for these are defined by upward and downward motion.

But, since up and down and the other differences of place seem to exist relatively, one might raise the problem whether place itself is not a natural entity but has its being through its relation to us. For clearly the roof is above to the people in a house, but is below to those on the tiles. So if place should exist relatively to us it is clear that it neither has any power, as existing relative to our position, nor would its treatment be necessary for the natural scientist. In removing this objection he shows that up and down are natural and not merely relative. He proves it as follows: natural up and down are invariant. For above is whither light things move locally and below whither heavy things move. But things that are relative to us are not always the same, but, as we turn about, so they too change their position. For the pillar now on my right becomes on my left when I turn round or move, and the case of the roof is the same, as said above. So the natural above and below are not relative to us, nor is their existence relational; they do not differ only by their position relative to us but also in their power. Otherwise they would not have been receptive of different things nor have determined the specific nature of what went their way.

Aristotle also proves from mathematical objects the fact that differences of place are not only because of a relation to us but are natural, and that natural differences and differences of position are distinct. For if objects that have a left and right and up and down only by their relation to us are not in a place according to their own nature, as is the case with mathematical objects, those things that are in a place cannot have their differences of place merely by their relation to us. Aristotle seems to be only supposing the conditional when he says 'though not in a place, nevertheless they have a right and left by their relation to us' (208b23-4), as if he had said 'mathematical objects have a right and left only through their relation to us because they are not naturally in a place, since, if they were, they would have had these differences not merely by position but by their nature'.[18] Proceeding, he again makes the point by saying of their differences of place that 'they have them solely from position and do not have any of them naturally' (208b24-5).

It is clear from this argument that relative place is one thing and natural place another, and that relative place is capable of holding

[17] Thus one's head is at the top of one's body, whatever one's posture. (J.O.U.)
[18] Extending the quotation from Diels's 526,2 to 526,4.

things not in a place through our way of looking at them – for this diagram has a right and left relative to me as I draw it, and an above and below, as do its parts likewise. But natural place holds only for things that are in a place. Now all natural bodies and artifacts are in a place, but artifacts are only contingently in a place *qua* artifacts, but essentially so *qua* natural. Among natural bodies, each of the simple ones has its own proper place, but the proper place of a composite object is determined by which of the simples in them is dominant.

Alexander rewrites the passage that runs 'they have them solely from position and do not have any of them naturally' (208b24-5) to read 'so that their position is merely conceptual'. He says: 'For having declared that mathematical objects have a right and left according to their position relative to us, he adds what sort of position it is – that it is according to our way of looking at them. For if they do not even exist independently they could not independently have a position.' Here Alexander clearly accepts that the illustration from mathematical objects is not a separate proof, but is used as a confirmation that differences of place are naturally determined, since those things that do not have naturally determined places, such as mathematical objects, are definitely not in a place. But perhaps this could be an independent argument, as Alexander had earlier thought, when he said that the third of the arguments demonstrating the existence of place was that from mathematical objects. This might be so. But if mathematical objects, though not naturally in a place, none the less exhibit differences of place through their position, it is clear that place is something; for what is so through position is derived from what is natural, as things imagined are from things perceived.

> **208b25-9** Further, those who affirm the existence of the void [call it place; for the void would be a place deprived of body. That place is something in addition to bodies and that every perceptible body is in a place] can be grasped from these considerations.

Aristotle proposes this third argument to show that place exists of
527,1 its own nature. For those who affirm the void say that it is nothing other than place deprived of body. If, then, it is something in addition to body, so that it can exist without body, it is clear that it has some nature of its own. But he does not put forward the argument as acceptable to himself, but as establishing that earlier philosophers also had a conception of place, because, he says, 'those who affirm the existence of the void call it place'. They are accustomed to hold that place exists and has a distinct nature that is

perhaps superior to that of body, since body cannot exist without place but place exists also in the absence of body. He has proved the existence of place demonstratively by the two earlier arguments and concludes the discussion with the present argument based on opinions. He has proved that place exists, that it is distinct from bodies, that it has its own proper nature and power, and that it is not to be regarded from our relative position. He says 'every perceptible body is in a place' because of mathematical bodies which do not exist essentially in a place.

> **208b29-209a2** Hesiod, too, might seem to have spoken truly [when he made the gap (*khaos*) first. At least, he says 'Chaos came to be first of all things, but then the broad-breasted earth' because there must be a space for things first, since he thought, like the majority, that all things are somewhere and in a place. If it is like this, place must have a remarkable power and be prior to all things. For that without which nothing else exists, but can exist without other things must of necessity be first.] For place does not cease to exist when the things in it are destroyed.

Hesiod says in the *Theogony*[19]

> Chaos came into existence first.

The word '*khaos*' seems to be derived from the verb '*khô*' (to pour) and to imply, through that, accommodation (*khorêtikon*), which is the same thing as providing a place. From this Aristotle draws two conclusions: that place is something according to Hesiod and that according to him its power is prior to all others and originative and therefore appropriate in a discussion of first principles. He says that the reason why Hesiod said that chaos came first is that he thought, like the majority, who are ignorant of the incorporeal nature and believe that bodies alone exist, that everything is somewhere and in a place. He thought that, this being so, there must already exist a space in which things will be. By saying 'like the majority' Aristotle indicated that he did not accept the belief. He offered as an explanation of Hesiod seeming to be right the claim that it was necessary for a space for things to exist first. But if Hesiod's view that a space or place came into being first is true, the nature of place should be something remarkable as originally creative and originative. For if (1) it is not necessary that body should exist when place exists, but (2) it is necessary that place should exist when body

[19] Hesiod *Theogony* 116. (J.O.U.)

exists, if it is necessary that everything should be somewhere and in a place and that what takes other things with it if destroyed, but not vice versa, is prior in nature, then it is clear that place should be originally creative and originative. For if a particular body such as fire or air should be destroyed it is not necessary for place to be destroyed with it. But if place were a void, as some used to say, it would be even more necessary that it should not be destroyed together with bodies, since the void is a place deprived of body. The opinion of Aristotle about place holds that place is the limit of the container,[20] as we shall learn, so that, even if place is not destroyed with what is in it, it is still necessary, if place exists, that body should exist, and, if not always the same body, still invariably some body. But if place came into being first, and if there were a void, it is clear that place would exist even if body did not. These common opinions are stated as useful to show that place, like the unlimited, seems to some to count as a first principle. For this reason also the discussion of place is necessary for anyone who proposes to examine first principles.

It seems that then a type of exegesis of Hesiod's verses prevailed that linked 'chaos' with 'space' (*khôra*). But it does not signify space but the limitless and manifold cause of the gods, which Orpheus called 'the vast gap'. For after the one cause of all things that Orpheus proclaims as time, as being the measure of the mythical birth of the gods, he says that the aether and the vast gap proceeded, the one as the cause of the formally limited procession[21] of the gods, the other of the formally unlimited. Of it he says

no limit or base or seat sustained it.

But how could anyone say these things of place that bounds other things and supports them? But Hesiod, honouring the first cause of all things in silence, only hinted at it darkly by saying that chaos came into being; for it is necessary for everything that comes into being to be brought about by some cause, as Plato says. Wishing to show forth the manifold procession of the gods through the parallel case of the boundless, he presented the theogony by naming first chaos and then the earth.

209a2-4 But still there is a problem what it is, if it is, [whether it is the bulk of a body or of a different nature.] For we must first seek for its genus.

[20] i.e. the inner surface of what surrounds.

[21] 'Procession', like 'proceed', is a translation of the technical Neoplatonic term for the coming into being of derivative things from first or prior kinds of being. (J.O.U.)

In his discussion of the unlimited he set out five arguments by which the unlimited seems to exist and brought against them those through which it will appear not to exist, in order that the arbitrator of the arguments, who finds both true though appearing to contradict each other, may judge that in one way the unlimited does, in another way does not exist, which is the truth. Here he does not wish to show that place in one way is and in another is not, but that it is, simply through the common conception about it; for who does not say that place exists? So he did not immediately set up opposing arguments to show that it does not exist, but, having stated the arguments by which place is believed in, he seems to move on to the next problem of the nature of place. He shows that it cannot be either corporeal or incorporeal, nor an element, nor composed of elements. He brings forward other arguments also, through which he concluded that one must raise the question not only of the nature of place but whether it exists at all. In this way he does not propose that in a way it exists and in a way it does not, but indicates the difficulty of the discussion because of the puzzle common both to the issue of its nature and to that of its existence. For if it is necessary that, if it exists completely, it is either a body, or incorporeal, or an element, or composed of elements, but can be none of these, both its nature and its existence will appear problematic. Also the difficulty of discovering it makes the arguments in favour of its non-existence more persuasive, even if we all seem to have a concept of the existence of place. Observe how methodically he employed his argument. Wishing to exhibit the difficulty of the enquiry through all the problems, on the question of its existence he did not confirm it by the common preconception of it, but skilfully threw doubt on it by introducing it by means of the hypothesis of the void and the mythical chaos. On the topic of the nature and character of place he even does away with its existence. That his arguments tend in this direction is shown by his now speaking hypothetically of its existence.

Since the existence of place has not yet been completely proved, even if some arguments were offered in support of it, proof will be provided as the discussion proceeds. At the conclusion of the discussion of its character he clearly said that its existence was problematic; but first he inquires to what genus place should be assigned, for the genus is fundamental with regard both to existence and to definition.

So is place corporeal or incorporeal? He reasonably did not say 'or incorporeal' but 'or of some other nature', since he does not set it completely apart from body, for he says that it is the limit of a body, as we shall learn.

209a4-7 So it has three dimensions[22] [length, breadth and depth, by which all body is defined. But it is impossible for place to be body;] for two bodies would be in the same place.

Alexander distinguishes two syllogistic arguments opposed to each other in this quotation. 'The first of these is as follows: place has three dimensions; but what has three dimensions is body; so place is body.' But if every body has three dimensions, not everything having three dimensions is *ipso facto* body. For those who assert the existence of the void say that it is three-dimensional. 'The second syllogistic argument, opposed to the first, is hypothetical denying the consequent, which is also called the second indemonstrable. If place is body, two bodies will coincide – place and what is in it. But it is impossible for two bodies to coincide. So place is not body. But if it seems to be body but cannot be body it cannot exist at all.'[23] That is what Alexander says. But perhaps the refutation of the existence of place is constructed by division. For if it altogether exists it is in the genus either of the corporeal or of the incorporeal. But it is not incorporeal since it has length, breadth and depth, which are defining properties of all body; nor is it body, lest two bodies should be in the same place. So Alexander has well displayed the two segments of the division.

The ancient philosophers assumed that it was obviously absurd that one body should permeate another.[24] But the Stoics later accepted it as following from their own hypotheses, which they thought had to be supported in every way. For, holding it right to say that everything was corporeal, including qualities and the soul, and seeing the soul and qualities as permeating the whole body, they accepted that bodies permeated bodies in mixtures. Alexander shows that this is impossible both in a separate work[25] and in his commentaries by means of many arguments. First, if body permeates body, and everything permeates everything, if this is *qua* bodies, and they do not permeate each other *qua* hot or cold or as having some other difference, then, if body permeates body, the

[22] Aristotle's definition of place as a container appears to make place two-dimensional. Did he realise this? Here he firmly states that it is three-dimensional. He does not solve the problems raised by taking place to be two-dimensional, nor does he anywhere explicitly say that it is. (J.O.U.)

[23] In traditional logic hypothetical arguments have two valid forms. One is 'If p, then q, but p, therefore q.' This is hypothetical argument by affirmation of the antecedent. The other is 'If p then q, but not q, therefore not p.' This is hypothetical argument by denial of the consequent, as in the text. Argument by affirmation of the consequent or denial of the antecedent would be invalid. (J.O.U.)

[24] See Sorabji, *Matter, Space and Motion*, chs. 5-7, esp. ch. 6 for the Stoics, which rejects the interpretation of Stoic qualities and soul offered here, but agrees the Stoics allowed bodies to permeate each other in other contexts.

[25] Alexander *de Mixtu*, translated into English by Robert Todd (Leiden, 1976).

greatest will permeate the smallest.[26] For it will be unchanged before and after reception. For it did not then contain certain empty areas and channels but no longer now, since those who say that bodies permeate each other do not ascribe it to channels. So a body can receive a second and a third in the same way. In the same way, says Aristotle, the universe will be contained in a grain and the sea in a wine ladle.

Further, if body permeates body and the less solid the more solid, why does gold receive bronze but not water?[27] It is also inexplicable why gold, on receiving the bronze, has acquired a greater bulk, and in general why some of the things mingled will increase, if things equal to each other spread out alongside to surround the whole of each other so as to have the same surface.[28] For when a ladle is mingled with a ladle the whole ought to be a ladle in volume if they permeate each other.

Eudemus also adds this point, that it is not possible for bodies to coincide whether both are perceptible or if one is mathematical. For there will not only be two coincident bodies but also the mathematical one will be somewhere, which is implausible. But if the sphere of each of the heavens is complete, extending right to the centre, it is clear that the heavens will permeate each other and the stars and the whole of the sublunar. So perhaps sublunar bodies do not permeate each other, as being most material, resistant and naturally separated, which is why when mingled they form something bigger, while celestial bodies are different. So the hypothesis of the heavens and the grain is also absurd, since the heaven is indivisible. But if that which has three dimensions is not body an explanation is needed; for it is not easy to believe that there is something incorporeal that is three-dimensional.

> **209a7-13** Further,[29] if there is a place and space of bodies, [it is clear that surfaces and other limits will have them also; for the same argument will apply. For where there were formerly the surfaces of the water there will be instead those of the air. But we have no difference between a point and the place of a point, so that if place is nothing different from the point, neither will

[26] The argument is that if you put a cup of sea water into a cup full of wine, there'll be no spillage, if body can permeate body, so you can do it again as many times as you like, until you have put the sea into a cup full of wine, or the universe into a grain. Sorabji argues that the Stoics do not need to deny that the wine would expand and spill, though they may lack a further explanation for the fact. See Sorabji, *Matter, Space and Motion*, 100.

[27] Replacing a full stop by a question mark. (J.O.U.)

[28] Replacing a question mark by a full stop. (J.O.U.)

[29] *eti eiper* in Ross; Simplicius has *eti de eiper*. (J.O.U.)

> it be from any of the others,] nor is place anything beyond each of these.

This argument also seems to be concerned with definition, to the effect that place is not the space of a body. But he completely refutes the existence of place by showing that there is no such thing as the place or space of a body. The whole trend of the argument is as follows: if there is a place or space of a body, its boundaries will have them too and so, also, surfaces and edges and points. But, as he will prove, there is no place of a point; therefore there is no place of a body. He also shows that the conclusion is true by the argument introducing the very notion of place. For if we say that where there was formerly water there can come to be air and that place is something that is different from what is in it, it is clear that where there was formerly the surface of the water there is now that of the air. For the boundaries of each are translated with what they bound. The same is true of the edge and the point, so that there should be a place of a point by the same argument as there is a place of a body. But one could show that there is no place of a point and that we have no way of distinguishing a point and its place as follows: if a place is equal with what is in it, then there will be a place of a point having no parts; but what has no parts is a point; so the place will be a point of a point. But two points, if coincident, become in actuality one point and are no longer two. So a point is not one thing and the place of the point another. There would have to have been a difference between place and what is in it if place is something. Similarly there will not be a place of an edge or a surface, so by making use of the argument from exchange of position we previously proved the existence of place, and now its non-existence.

532,1 **209a13-18** For what would we posit place to be? [It cannot be an element or composed of elements, whether corporeal or incorporeal, if it has the required nature. For it has size, but no body; but the elements of perceptible things are bodies,] while no magnitude is composed of intelligible elements.

Since all things that exist are either bodies or incorporeal and either elements or composed of elements, i.e. either simple or compound, having shown that place is neither body nor incorporeal, he now demonstrates the remaining conclusion that place is neither an element nor composed of elements, if it has the required nature, so as to have magnitude, if it is equal to what is in it, but not to be a body, if two bodies cannot be in the same place. For by not being a body it could not be among perceptible and corporeal things, whether composed of elements or as an element. For the elements of compound bodies are also bodies. For what is proximate to compound bodies is either fire and air and water and earth, or atoms, according to Democritus, or homeomeries,[30] according to Anaxagoras. But nor can it be among intelligible things, whether as composed of elements or as an element. For intelligible elements do not have a magnitude, nor could something having magnitude come to be from intelligible elements without magnitude. So if place has a magnitude it could not be composed of intelligible elements or be an intelligible element. So it could not exist at all, if everything that exists is either an element or composed of elements.

It is worth considering what he meant by saying that the elements of perceptible things were bodies. For matter and form are the primary elements of perceptible things, and they are both incorporeal. Perhaps, then, he did not say that the elements of perceptible things were bodies pointlessly, but in order to indicate what sort of compounds he was referring to – not those immediately consisting of form and matter, but those composed of the four simple corporeal elements, which are not in themselves perceptible. Nor do they exist absolutely in themselves; for the sense organs themselves are not simple but compound. Earth itself and water and air and fire are perceptible because they are not simple,[31] but are so called and exist because of the predominance of the element. Also it seems to me that Aristotle was aware that he was not bringing about the abolition of all elements, since matter and form were unaffected, but only of bodily elements, and that is why he introduced the argument that follows, in which he will show that place is not a principle or element as being matter or form.

> **209a18-22** Further, of what in things would one posit that place was the cause? [For none of the four causes belongs to it. It is not a cause as the matter of things, since nothing is composed of it, nor as form and account of things, nor as an end,] nor does it change things.

To demonstrate that place does not serve as matter or as form is to fill up the omission of the preceding theorem in which he showed that place is not an element whether of perceptible or of intelligible things. He had previously shown that it is not a perceptible element
in perceptible things. Now he shows that it does not serve as matter 533,1
or form. But since matter and form are also causes, he makes the remainder of the demonstration general, that place is none of the causes, neither as matter nor as form nor as efficient nor as end. He

[30] Homogeneous stuffs.
[31] Any actual sample of an element is contaminated with the other three.

showed that it is not an element or principle in general nor a cause of things as matter in this way: matter is that of which a thing is composed; for material things are composed of matter; but place is something of which nothing is composed; so place is not matter. He himself omitted to show that place is not form nor final nor efficient cause. One might show that it is not form by the fact that place is predicated of other things as well as of what is in place. Also in exchanges of position the forms are changed but the place is not. Nor is place an efficient cause. For it receives what already has come to be and does not bring about what is not. Moreover, place does not change or alter what is in it, but the reverse, at least if it makes it rest. For what has come into its own proper place is at rest. Therefore Eudemus in the third book of his *Physics*, paralleling the present passage and stripping all of the causes from place as something accepted, says: 'Is it then the cause of change? But that also is impossible, Democritus; for it needs to have some motive power.' For Democritus said that place was the void, and as void in its own nature it would be powerless and inactive.

Perhaps then place is the final cause. This seems plausible because of the motion of bodies towards it and the aim of bodies to attain their own proper place, in which on arrival bodies remain at rest. And the special feature of an end is that what achieves it remains at rest. However, if place is one thing and being in a place another, as is shown by the category *where* (place) being other than place itself, which is assigned to the category of *how much* (quantity), and if the end of bodies is, if anything, to be in this place, then place would not be the end.[32] But in general the end and the goal is not something in itself, but is the completion of that of which it is said to be the end. For in that way happiness and likeness to god are the end as being present in the happy and godlike man, not as self-subsistent, as place is said to be. This must specially be said, that if place *qua* place is an end, and if whence water has departed there comes to be air, there will be the same end for different bodies. This matter still required examination, since the commentators have passed it over in an off-hand fashion.

This argument contributes to the conclusion that place is nothing existent, whether as a principle or derived from a principle, and therefore to its absolute non-existence, if everything that is is either a principle or derived from one. It especially leads to the view that it is
534,1 not a principle as those who adduced Hesiod's chaos and the void of Democritus and place as the cause of the natural motion of bodies seemed to be saying.

[32] Being *in* a place would rather be the final cause or end, i.e. the category of *where*. Place itself is not a category, but falls under the category of quantity, although modern textbooks often overlook this.

209a23-5 Further, if place is among existents where will it be. [For Zeno's puzzle requires some response. For if everything that is is in some place, it is clear that there will be a place of place,] and so on without end.

This argument also is destructive of place existing at all. He introduces the thesis that everything that is is somewhere, since that which is nowhere does not exist at all, and he introduces it using Zeno's puzzle. For if everything that is is somewhere, if place is among existents it should be somewhere. But somewhere is in a place. Therefore place is in a place. And so there will be a place of place and a place of that place and so on without end, which is impossible. If, then, an impossibility followed from place being among existents, it is clear that place does not exist. So, unless Zeno's paradox were solved (which is what is meant by its requiring some response), it is clear that the being of place is refuted.

209a26-30 Further, just as every body is in a place, [so a body will be in every place; how then shall we speak of things that increase in size? For on this basis it will be necessary for place to increase with them, if the place of each is neither less nor greater than it. For these reasons we must raise the problem not only what it is] but also if it exists.

If place exists, but the void does not exist, as will be demonstrated, it will follow on every body being in a place that a body is in every place. If this is true it is necessary that place should equal what is in the place. Since, then, it will be necessary for the place of growing bodies to grow with them, what is the growth of a place? How and whence does it come about? For what comes to be comes to be from its opposite. He plausibly assumed that there was a body in every place; for if there were a void, that which grows would grow into it. It is clear that either body comes to be and passes away – another thing comes to be as one passes away – or it remains the same and there will be no need for place to come to be. This argument also starts from the question of what place is, demonstrating that it is not the receptacle[33] of body and showing at the same time that place does not exist at all. On the basis of these arguments he reasonably concludes that 'for these reasons we must raise not only the problem what it is but if it exists at all'. For if it is neither body nor incorporeal, neither an element nor composed of elements, not a principle or cause, nor equal to what is in place, which seems to be

[33] It is Plato's account of space (*khôra*) in the *Timaeus* that it is the receptacle which receives qualities or bodies.

the case, it is clear that both the nature of place and its existence must be problematic.

CHAPTER 2

535,1 **209a31-b6** Something may be predicated of a subject either in virtue of its own nature or in virtue of something else. [Thus there is the place that is common, in which all bodies are, and that which is private to a thing and primary. I mean, for example, that you are now in the heavens because you are in the air and that is in the heavens, and you are in the air because you are on the earth, and likewise on the earth because you are in this particular place that contains nothing but you. So if place is that which primarily contains each body, it should be a sort of limit,] and thus place might seem to be the form and shape of each thing, by which its magnitude and the matter of the magnitude is determined. For this is the limit of each thing.

In our problems about each subject, after the questions whether it exists and its definition, we raise the question of what it is like. But we come to know what a thing is like from its essential nature. In the case of place also, after the questions if it exists and of its definition, he moves on to the question of what it is like. In considering the attributes of place he discovers here also the difficulty of the problem, since some of the attributes ascribed to place lead to the thought that place is form, others to the thought that it exists as matter. That is why in drawing a conclusion he added that 'as we examine the question on this basis it may reasonably seem difficult to discover the nature of place'.

Since, then, it seems to be an attribute of place that it is the prime container, from this he shows that place is virtually form, arguing as follows: the private place of each thing is its prime container; the prime container is the thing's own limit; but this is its form. That place taken strictly is the primary container he showed from the fact that a certain place is spoken of as common, in which all or many bodies are, as in the universe (he now says 'the heavens' instead of 'the universe') or in the air or on earth; I am in one of these as in a common place because other bodies are in it also. But when I am said to be on earth, or at home, as being in this particular place, in that it contains nothing but me, then that is my private place, in which I am primarily contained. The private place is the one that strictly contains me. But since he wished to distinguish the immediate and primary container from the common and less immediate place, predicated with external reference, he first spoke generally, saying that some things are spoken of with reference to their own nature and some with reference to something else, saying 'in virtue of their own nature' instead of 'primarily'. Basically 'of its own nature' is opposed to 'contingently' and 'with reference to something else' to 'primarily'. Things are said to obtain of their own nature when they are part of the essence of those things in which they are said to obtain, like number in the even and odd; in general genera obtain of their own nature in species because they are partially constitutive of their essence and belong to them as such. Species also belong to genera of their own nature, since they are a subdivision of the essence of the genera. *Qua* number, this one is even, that one is odd. What is primary exhibits immediacy. Sometimes these go together, so that obtaining of its own nature and obtaining primarily are the same thing, as animal obtains of man, sometimes they are disparate, so as to be of their own nature but not primarily. In this way substance obtains of a man of its own nature and quantity of the even; for not only genera are predicated of species of their own nature, but also the genera of genera, yet not primarily nor immediately but through the medium of the immediate genera. But a surface is white primarily, but not of its own nature, but contingently.

These matters, however, no doubt belong to the study of logic. But that he here said 'of its own nature' instead of 'primarily' is clear from the antithesis to 'with reference to something else', and he himself makes it clear by changing soon after to 'primarily', when he says that the common place is that in which all bodies are, but the private that in which a thing primarily is. Generally, he now needs the concept of immediacy in particular. But the immediate is the primary but not what is of its own nature. But perhaps he did not call the private and equal place both of its own nature and primary casually, since this place is both primary and of its own nature, whereas the more voluminous place is contingent. For I am at home or in the city – because the place in which I primarily am is contingently in my home and my home in the city. But the equal and like-shaped place is mine also of its own nature.

But since each thing is bounded by that which contains it immediately and primarily, it follows that what primarily contains a thing is its limit. So the primary and immediate container of a thing should be its private place; so place is a limit. But what contains a thing immediately and is its limit is in each case its form. For the form is what limits and bounds each body. Therefore the individual form of each thing which limits it as its shape is also place.

Further, he said 'by which the magnitude is determined' rather than 'the material interval' because magnitude is of two sorts. One of these is indeterminate and material, viewed as extension and diffusion, the other is determinate and formal and bounded according to shape and measure. This is why he added 'and the matter of the magnitude', not as saying something beyond what had already been said, but as making it clearer that also previously the determined magnitude was the material element in magnitude.

Further he said 'in the air because it is also on the earth' rather than 'in the air surrounding the earth'; for through this it is in the whole air. And reasonably the proper place of each thing is that which surrounds it and is private to it. Through this it is in the more voluminous place. You are in the house because you are in this place in it, which fits you, and in the city because you are in the house, and on the earth because you are in the city. So if bodies are limited by their immediate container and by this primarily (for they are not limited by anything else before it nor is there anything in between), the form should be the limit of each thing, because of one of the attributes of place, which is to contain immediately. In another respect place will appear to be matter, as is clear from what follows. The paralogism[34] occurs because of the ambiguity of 'immediate'. For of immediate containers one is intrinsic, the other extrinsic, like place.

> **209b6-11** In the way that place seems to be the interval of the magnitude, [it seems to be matter. For this is other than the magnitude; it is what is contained by the form and determined by it, as by a plane and limit. Matter and the indeterminate is like that; for, when the boundary and the properties of the sphere are abstracted from,] nothing is left other than the matter.

Having stated one of the attributes of place, with regard to which place seems to be the form of bodies, he now speaks of another of its attributes, with regard to which place will no longer appear to be the form but to be the matter of what is in place. For, he says, if place seems to be the interval of the magnitude, place will be no longer the form but the matter. Now place seems to be the interval receptive of the magnitude; for a magnitude that already has a form is said to be in a place. Within it there is, first, an interval, indeterminate after the abstraction of the limit and the other attributes which it receives when viewed privatively, but also there is the limit and boundary determining and containing the indeterminate interval, and this is the form. After abstraction from this there is nothing left over

[34] Of 535,13ff. that place is virtually form.

except the matter, i.e. the interval. If, then, place is the interval of the magnitude and the interval of the magnitude is the matter, the consequence is clear. But it is remarkable that he likened the matter to place *qua* interval and not *qua* receptacle. But perhaps he deferred here to Plato, as we shall learn. Or, rather, he viewed an interval as receptive.

But if the interval of magnitude is a quantity and is measurable, and if quantity is a form, how can the interval of a magnitude be said to be matter? Perhaps the material interval is not a measured quantity nor a form, unless it partakes of the form of quantity and magnitude, but is a formless diffusion. Aristotle clearly has this conception of it where he says that 'such a thing is the matter and the indeterminate'. Also, when he says 'in the respect that place seems to be the interval of the magnitude', in that respect place is the matter, he is calling the indeterminate substrate the interval of magnitude, which he had already called unlimited. It is as this that he also called the interval the matter of the completion of magnitude.

So let nobody think that bodily extension as a quantified magnitude or as the numerically determined distinction of the manifold belongs to bodies from their matter, but only the scattered diffusion and indeterminacy in them, through which material forms differ from the immaterial. This seems to me to be specially appropriate to correct views about matter. But Alexander, although he said much towards the distinction of material extension from magnitude, seems none the less to be disturbed by it on the ground that every quantified extension of magnitude is formal. So he was led to comment on the passage that 'we should understand what is here said about matter as worded not accurately but according to the requirements of the projected demonstration'. However, as I said, Aristotle has clearly distinguished material extension from magnitude by it being indeterminate and unbounded, and he has plainly said that this interval is other than magnitude and is the matter of the completeness of magnitude and is contained by the form as by a surface and limit and is indeterminate in its own nature. For, as I have said, one must not think of the material extension as magnitude but as a formless scattering[35] of the formal simplicity and connectedness.

But Alexander, having enquired what form it is that determines the matter, says: 'Aristotle made it clear when he said "as by a surface and limit". For the form, taken in the sense of shape, is that which determines and circumscribes matter and limits it, not the form taken as potentiality and account. For if[36] lead is defined by its weight and this is its form, this, however, is not its form *qua* body, but its

[35] Reading e.g. *diaspasmon* for *parainesin*. (J.O.U.)
[36] Reading *ei gar* with a for the *ou gar* of EF. (J.O.U.)

being bounded by a surface.' That is Alexander's view. But perhaps everything, including the qualities, should be taken as the form determining the matter. For that seems to be Aristotle's idea when he says that 'when the limit and the attributes of the sphere are abstracted from, nothing is left except the matter'. If one should wish to discover the form of magnitude *qua* magnitude compounded of matter and form, he will not find the surfaces and the limit containing it from the outside. For these are the limit of the form, but not the form, and the surface, but not the solid. If, hypothetically, there was an unlimited body, the form of the body would be extended everywhere, but it would not be bounded by surfaces. For the form of the continuous quantity as such is not its surface, but its whole nature that permeates it. If he himself in the previous book says that anything quantitative is also a particular quantity (for we read the term '*poson*' as having these two meanings),[37] the particular quantity will be determinate. For what is not one of the particular quantities is simply not a quantity at all. This is true, if the general has its existence in particular things, but still the form of the continuous quantity is seen in the general also, existing no less because it is common to determinate things.
539,1 However, Aristotle does not seem to assign to the form only circumscription by surfaces, but mentioned what was conspicuous as an example (which is why he added the word 'as' in 'as by a surface and a limit'). Continuing, he said more accurately: 'for when the limit and the attributes of the sphere are abstracted from, nothing remains except the matter.'

> **209b11-17** That is why Plato in the *Timaeus* says that matter[38] and space [are the same thing. For the participant[39] and space are one and the same. There he speaks differently of the participant from in what are called his unwritten doctrines, but he still declared that place and space were identical. For everyone says that place is something, but Plato] alone tried to say what it is.

Having said that in so far as it is an attribute of place to be an extension we shall be led back to matter in our enquiry into the nature of place, he commented on Plato having called matter the space and place of embodied forms in the *Timaeus*. For in the

[37] Literally 'for we read "*poson*" as having first an oxytone and then a barytone accent'. (J.O.U.)

[38] It is in fact Aristotle who identifies the space of Plato's *Timaeus* with matter – a word which Aristotle introduced, and a concept which he first isolated.

[39] Space participates in fire-like and other qualities derived from the Forms in Plato's account.

Timaeus[40] he says that matter is the receptacle of all becoming like a nurse. As he continues, he makes clearer the content of what he says. There is, perhaps, nothing against quoting more of what Plato had to say: 'these things being so, it must be agreed that one kind of being is the form that is always the same, having no beginning and indestructible, receiving into itself nothing else from elsewhere, nor entering into anything else, invisible and in other ways imperceptible, the observation of which is assigned to intelligence. Second is that which is like-named and similar to it, perceptible, created, set in motion, continually coming into being in some place and again passing away from it, grasped by opinion with perception. Third there is the kind of being that is space which also is indestructible, providing a resting-place to everything that comes to be; it is grasped without perception by a bastard form of reasoning and is unreliable; we regard it in a dream and say that it is necessary for everything that is to be somewhere and occupy some place and that that which is neither on earth nor in the heavens is nothing. Of all these things, and those akin to them that surround the undreamed and true nature, we are unable to speak with accurate judgment and awake because of our dreamy state. For it is fitting that an image, since it does not own even that which comes from its prototype but is always the changing shadow of something else, should therefore come to be in something else, clinging to being in some way or other, or be nothing whatsoever. The true and accurate account that is a support of that which truly is holds that if two things are different then neither can come into the other and remain itself so that they are still two. So this account, worked out according to my judgment, is my concise offering, that being, 540,1 becoming and space are three separate things, that precede the heavens.'

Plato says these and similar things, and Aristotle here also looks to the apparent meaning of the text and says that Plato states that matter and space are the same thing, as do those who make the interval their starting point. Among these things that he seems incidentally to blame Plato for, he did not blame him for calling place the matter of the forms. For he knew himself that one thing was referred to as the place that received the forms when he said in *On the Soul* that 'those who say that the soul is the place of the forms speak well, save that the forms are there potentially and not in actuality',[41] and something else was referred to as the place that receives body, which is what the present discussion is investigating. The only thing he says against Plato is this: if he says that place is

[40] See Plato *Timaeus* 51E-52D. (J.O.U.)

[41] An abbreviated paraphrase of *de Anima* 429a27. (J.O.U.)

matter, but the forms and numbers are also in matter according to him, since these also partake in matter, how can he say that the forms are not in a place? For that is why there is no place beyond the heavens, as he said in the discussion of the boundless.[42] But perhaps even this is not a criticism of Plato, but he is pointing out to those who can understand that Plato did not say that matter was place in the sense in which 'place' is used in the present investigation. For he would not have denied that the forms and numbers were in a place, if he wanted matter to partake of them. That, then, is what Aristotle says. Alexander agrees also that matter is called space in the *Timaeus* with a different meaning, but he says that Aristotle finds fault with Plato with good reason. For even if someone says that he calls matter place metaphorically, anyone speaking metaphorically ought first to have it established what place is in the basic sense. But if he is found nowhere else discussing place he seems to Alexander to be worthy of censure, because he treats place metaphorically without having first discussed place in the basic sense.

But it is clear from the passage quoted from him above that Plato calls matter the place of forms and not of bodies. It is also clear from what has been said that such a usage is not simply metaphorical, but is taken according to some common significance of 'place' as that which is receptive, just as Aristotle also thinks it right to call the soul the place of forms. For if what partakes of something and is determined by it receives that of which it partakes, and if what receives and accommodates it becomes the space of what is in it, and if a space is thought of as place and if matter partakes of the forms,
541,1 then place will be of the forms but not in the way that it is of bodies. The place of bodies is different. Here he treated what partakes as place.

But Plato seems to know another significance of place more embracing than that as what determines proper ordering. Whether this comes about through position, as in the case of bodies, or through participation, as in the case of things potential and actual, or through mere determination of order, it seems to be place. That is why an order seems to be according to some position; for what is assigned to a certain order seems to be situated in it, and everything is situated in a place. In accordance with this general significance of place as what determines an order Plato even speaks of a supercelestial place when he says in the *Phaedrus*[43] 'But the supercelestial place no poet has ever sung worthily, nor will he ever'. So what is said about matter as being about place should be taken as

[42] Aristotle *de Caelo* 1.9, 279a11-18.
[43] Plato *Phaedrus* 247C.

of the place that partakes not of bodies but of forms. But I know that Plato never researched seriously into the nature of the place of bodies nor made his views public. For how could matter be the place of bodies, if bodies are composites, whether of matter and form or as form in matter? For in that case matter would be receptive not only of form but also of matter. In any case if a body moves from place to place, but does not move from matter, matter could not be place. Moreover we shall soon hear Aristotle destroying that hypothesis. It is clear that different things are referred to as place in Plato, one with reference to matter, another spoken of in the *Phaedrus* with reference to the intelligible, as is shown by the fact that place with reference to matter is receptive and determining and makes one with what is within it, which is clearly denied of intelligible things, to which he assigned the supercelestial place in the *Phaedrus*. For having spoken of matter as of place and space and receptacle and seat of the forms, in saying how it is known he calls it 'grasped without perception' and 'by a bastard form of reasoning and unreliable'; he adds that 'we dream as we regard it' and the rest of what was quoted a little earlier.

But if Plato appears to have said nothing about place as that which receives bodies, why does Aristotle say that Plato alone had tried to say what place is? Is it because Plato seems to say about that place which is matter regarded as place that it is a receptacle, that it is eternal, that it is the seat of forms in becoming, that it is grasped without perception, unreliably by a bastard form of reasoning and so on, and Aristotle says that he has spoken of place in so far as he has called matter place and space? But perhaps he was aware that he
above all knew the depth of Plato's wisdom and that even if he did not 542,1
seem to make a separate inquiry into place he was contemplating it as well as bodies, since place has its existence conjoined to bodies. For he speaks in the *Timaeus*[44] about the power of place: 'Being affected in these ways everything changes its location. For the manifolds are sundered from each kind in their own place through the change of the receptacle that receives them, and some, becoming unlike themselves on each occasion and like to other things, are borne because of the shaking to the place of those things to which they have become like.' Aristotle says Plato spoke of matter in one way in the *Timaeus* and in another in his unwritten discussions. For in the *Timaeus* he calls it that which partakes[45] (for it partakes 'at the furthest remove from the intelligible'[46]), but in the unwritten discussions he called it the great and the small.[47]

[44] Plato *Timaeus* 57C. (J.O.U.)

[45] It partakes of qualities or forms, so as to take on appearances like fire, air, earth, or water.

[46] Plato *Timaeus* 51A. (J.O.U.)

[47] The great and the small, sometimes interpreted as extension of unspecified size,

That everyone agreed on the existence of place is clear from the fact that they all used the concept of difference of place.

> **209b17-21** If we examine the matter on this basis, [it not surprisingly seems difficult to find out what place is, if it has to be one of these two, either matter or form. These in any case need close scrutiny,] and are not easy to recognise apart.

Having said that some of the arguments about place suggest that place is form and shape, others that it is matter and substrate, he concludes from this also that the discussion of place is difficult. Also the discussion of matter is very difficult as such, if, according to Aristotle, matter is so unknowable as to be known by analogy only and, according to Plato, unreliably based on bastard reasoning. But knowledge of the form demands both a pure life[48] and exact thought. Also he who wishes to understand just one form needs knowledge of them all, because of the mutual sameness and difference of the forms. If both matter and form are basic principles, principles require the most close contemplation; yet it is impossible to recognise them apart. For they exist inseparable. For all these reasons it is hard to find out what place is, if it is either matter, or form, or both.

> **209b21-6** But it is not difficult to see that it is impossible for either of these to be place. [For the form and the matter are not divided from the thing, but it is possible for its place to be. For where there was air there is in turn water, as we said before, the water and the air exchanging places with each other, and other bodies likewise.][49]

Having stated the grounds on which one might suspect that place
543,1 was matter or form, he shows by seven arguments that it is impossible for place to be either of these. His first proof is a syllogism in the second figure[50] and is as follows: matter and form cannot be separated[51] from the thing to which they belong; place can be so separated; so place is neither matter nor form. It is very

---

plays a role like that of matter in one or more of Plato's unpublished lectures.

[48] Asceticism was considered an essential prerequisite by the Neoplatonists for intellectual understanding.

[49] No doubt through oversight, the lemma reads *heôs tou to de angeion ouden tou pragmatos estin*. As is clear from the text and the next lemma it should read *heôs tou allôn sômatôn homoiôs*. (J.O.U.)

[50] The second figure is of the form PM, SM, SP. In this case it is a syllogism in Cesare and is of the form No P is M, All S is M, No S is P. (J.O.U.)

[51] This argument turns on the *immobility* of place, which therefore does not accompany the body that leaves its place.

obvious that matter and form are inseparable from composite things so long as they exist. For the being of compounds consists in these. He inductively showed that place is separable while the thing in the place exists and is preserved because of exchange of place. 'For where there was air, there is in turn water.' That the argument is inductive he showed by adding 'and with other bodies likewise'.

> **209b27-8** So the place of each thing is not a portion nor a condition, but separate.[52]

Having proved that place is separate from the thing, he reasonably drew another conclusion, that place is also not a portion of what is in it. For if it were separated, being a portion, neither would it be any longer a portion when separated nor would that from which the portion was separated remain a whole. But nor can it be a condition, if separable. For a condition does not persist if separated, but ceases to be, while a place persists. One can interpret 'portion' as being matter and 'condition' as being form. But matter and attribute are portion and condition, as stated. A condition is a completion just as the form is. So place is neither matter nor form. But it is possible to interpret 'portion' more universally as including all that completes a thing's essence and interpret 'condition' as being a contingent attribute. For place is neither a quality, nor a power, nor any other portion of what is in place. Is place then not a contingent attribute of what is in place? Or is being in a place an inseparable attribute of a body, but the place not an attribute of what is in it, but, if at all, of that of which it is the limit and surface?

> **209b28-30** For place seems to be something like a vessel, [since a vessel is a moveable place. But the vessel] is no part of the thing in it.

The commentators tie this on to the second conclusion just stated. But he proves the same thing from the likeness of place to a vessel. For the vessel also holds what is in it, and he calls it a moveable place. Just as the vessel is no part of what is in it, so is place. But matter and form are part of the thing. There can be three 544,1
conclusions or there can be one. Of the other two, one can be taken as a support of the minor premiss, which says that place can be separated, the point about portion and condition being a support for the other premiss. Also there can be two arguments, the third stated being on its own, which is the most likely.

[52] As in the previous lemma the text reads *heôs tou to de angeion ouden tou pragmatos estin*. This is incorrect. The next lemma ends correctly with those words. (J.O.U.)

209b30-2 So, as being separable from the thing, [place is not the form, and, as containing it,] it is other than the matter.

Matter and form have it in common that they are inseparable from the composite, and from this he showed earlier jointly that place is neither matter nor form. But to be contained and not to contain is a special feature of matter that is different from that of place. So not wishing to omit this either he divided matter from place because matter is contained and place contains. He distinguishes form by making the former point again, that the place is separable and the form inseparable. But perhaps there is not now the same simple idea that the form is inseparable, as was the case before. For it would be excessive if twice thrown before us with the same purpose. But he indicated by the repetition that the form shares with place the character of containing, while matter differs from it in this way also. The form has the character of containment as containing inseparably with a shared nature, and not as separate like place. Alexander also makes a good point about the passage saying: 'Now from the predominant feature of each, the form and the matter, he lays hold on their difference from place. The predominant feature of the embodied form is that it is inseparable; for its separation from a thing is the end of that thing. But matter, even if it be inseparable from form, is still separable from this form. In the case of a man, for example, in the world of becoming, the form is destroyed when it is separated, but the substrate remains, taking on another form. So he reasonably showed that form is not place by its being inseparable, but that matter was not, since it is separable in one way, he showed, not from the composite thing, but from its being contained. For it is impossible for it to exist and prevail in what is unless determined by form.'

209b32-3 It seems that what is is always itself somewhere, and that there is always something else outside it.

What is said is clear, that both from what is manifest and from common understanding it is obvious that what is in place is one thing and the place another. The commentators treat this as being another, fourth, argument that the thing in place is one thing and place another because what the thing is in is outside it.[53] So if what is in place is matter and form, place must be something beyond them. But it could be a joint conclusion with those of preceding proofs based on obviousness. Such a view is more compatible with a text that contains the explanatory word 'for'; 'For it seems that what

[53] Reading *tou en hôi ektos ontos* with aF. (J.O.U.)

is is always somewhere', which occurs in some texts, seems to suggest this. For since what is in a place consisting in matter and form seems to be one thing, and place to be another, it is reasonable that place should be one thing and matter and form another.

> **209b33-210a2** If a digression is in order, Plato ought to tell [us why the forms and numbers are not in place, if that which participates is place, whether the participator be the great and the small or,] as he wrote in the *Timaeus*, matter.

In his discussion of the unlimited Aristotle said that Plato claimed that outside the heavens 'there is no body, nor the forms, since they are not anywhere'.[54] Here he recalls Plato describing the participator in forms as place in the sense of matter. It appears that these statements conflict. For if place is what participates in the forms, why are the forms, which he also called numbers, not in place? But in the unwritten lecture on the good he called the participator great and small, while in the *Timaeus* he called it matter, which he also called place and space. It is clear that this is an apparent contradiction, but it is not really contradictory. For he called matter the place of the embodied forms, as being their receptacle, but he said that the forms that are separate from matter, or the ideas, were not in a place. For these are not in matter as in a place as are embodied forms, nor in the receptacle of bodies as are bodies. The contradiction would have seemed more evident if somebody had adduced to him, when he said that ideas were neither in place nor beyond the heavens, the passage in the *Phaedrus*, where he says: 'But no poet has worthily praised the supercelestial 546,1
place, nor ever will. For the colourless and shapeless and intangible being that truly is and may be seen only by the governor of the soul and with which the true kind of knowledge is concerned occupies that place.'[55] In this place, he says, 'one sees justice itself, sees temperance, sees knowledge, not that to which becoming is present, nor that which is somewhere, being one thing in another and concerned with what we now call realities, but that which is knowledge in the realm of true being, and other true realities likewise.' For, observe, he clearly says that the ideas are in place, those of justice, and temperance and knowledge; for the idea is each of these itself. Also he says that this place is outside the heavens. But it is clear that it was one kind of place that he denied to things THERE, namely the sort of place that received enmattered forms or bodies, and another kind of place which he praised as determining

[54] Aristotle *Physics* 203a8. (J.O.U.)
[55] Plato *Phaedrus* 247C-D. (J.O.U.)

the order THERE.[56]

It is remarkable how Alexander, though he realises that Aristotle is now calling the ideas forms, none the less thinks that Plato is compelled to say that the ideas are in place, though he calls them immaterial, since he called matter the place and space of the embodied forms. But he calls this argument a digression, since, so far as what preceded it went, it would have sufficed to show that place is not matter, as Plato seemed to say; when he says that matter is not simply place but the place of the forms the opposite to what he says elsewhere follows – that the forms are in a place. Therefore he agrees that Aristotle digressed from his purpose in introducing this anomaly.

> **210a2-5** Further, how would anything move to its own place, [if place were matter or form? For it is impossible for what has no concern with motion nor with an above and below to be place.] So place must be sought among such things as do.

This is the fifth argument to prove that place is neither matter nor form, this also having the same course of argumentation: if place is either matter or form, nothing in motion would move towards its own place; but everything in motion moves towards its own place, heavy things downwards, light things upwards; so place is neither matter nor form. For a thing does not move towards its matter or its form, for both of these are with and in these moving things. For what is the moving thing other than matter and form?

Alexander says that the next statement that it is impossible that what has no concern with motion nor with an above and below to be place is another argument connected with that before it as follows: what does not admit motion is not place; but matter and form do not admit motion; so place is neither matter nor form. 'But', he says, 'what this contains may also be another argument as follows: what does not contain the distinctions of up and down is not place; but matter and form do not contain the distinctions of up and down; so place is neither matter nor form. But Aristotle', he says, 'put them together into one argument, adding the question how things in motion would move to their own place, if place were matter or form, taking it to be obvious that nothing moves towards its own form or matter.' That is the view of Alexander and of Aspasius. But perhaps the whole argument is one, as Alexander also noticed, and has the prior statement in conditional form as follows: if place is matter or form, nothing in motion would move towards its own place; but

[56] The capitalisation is intended to remind the reader that Simplicius uses this word to refer to the world of reality as opposed to the world of becoming, which he calls HERE. (J.O.U.)

everything in motion naturally moves towards its own place. In this he left out the further assumption as clear, but he proves the conditional by saying that it is impossible for what does not admit motion nor above and below to be place. For if motion cannot be towards matter or form, and if these have no above and below, towards which natural movements occur, there are features of matter and form different from those of place. From both these points it is clear that if place were either matter or form, nothing among things in motion would move towards its own place, that is, upwards or downwards. I am inclined to this reading of the passage also by the explanatory conjunction 'for'. For it is introduced as giving an explanation of why each thing would not move towards its own place if place were matter or form.

If the foregoing is true, one must not look for place in matter or in form, but in that which is distinguished by having an above and a below. For it is towards these that things in motion can move, not towards matter and form. It is possible thus categorically to pose the argument in the second figure of the syllogism: place is that towards which each thing in natural motion towards its own place moves; but nothing moves towards its own matter or form; therefore place is not form or matter. In general, if bodies always have matter and form and these are place, how could a body still move towards its own place?

> **210a5-9** If place is in itself, and it must be [if it is shape or matter, place will be in place. For both the form and the unlimited change and move about together with the thing, not always at the same place but where the thing is;] so there will be a place of place.

This I take to be the sixth argument to show that neither matter nor form is place, being also a *reductio ad absurdum*. It says that if place 548,1
is in the thing itself there will be a place of place; but this is impossible, both because of this very fact that place is of a body and not of a place, and because of the infinite regression if we continually posit a place of a place. So place is not in the thing itself. He strengthens the antecedent about this hypothetical place by 'it must be if it is matter or shape'. For if place is either matter or form, it must be in the thing itself which consists of matter and form. He shows that the consequent is sound by 'for both the form and the unlimited, i.e. the matter, change and move about together with the thing'. For as the composite thing changes, the form and the matter, of which the composite consists, change with it, and they are wherever the composite thing may be. If, then, place is these, it is clear that it will change as these change, and will change from place

to place; so place will be in place. It is clear from perception that natural bodies consisting of form and matter move and change their place. It is also clear that for places to move is absurd; for what moves must move from place to place.

Alexander raises a problem at this point. 'Perhaps it is not necessary, if something is in a place, that what is in it should also be in place, except contingently. For observe that the parts[57] of things in place and their qualities are not essentially in place, so that even if the composite be in place, it is not necessary for form or matter to be in a place, except contingently. So that even if form or matter were place, it would not be in a place, except contingently. So there is nothing absurd in place being in a place. For he himself, when he says that place is the limit of the container, says that the limit of the container and the contained are in the same place; for surfaces that touch each other are in the same.' As a solution of the problem he says: 'If the form or the matter is the place of the composite, it is equal to it; for a place is equal to what is in it. Since, then, the composite is equal to its place, where by motion it has come to be, and the things in it are equal to each other and to it, the place in which the composite body came to be would be equal to the form and to the matter. But what in a place is equal to the place is in place essentially and not contingently. For an entity is in a place by being extended. If, then, the matter or the form, being the place of the composite and coextended with it, is in the place where the composite is, and is equal to the place, it must be essentially in place. It is absurd', Alexander says, 'also that two places should be primarily occupied by the same thing
549,1 at one time, if both are equal to what is in them, both that in which the composite came to be and the form or the matter; for these are equal to the composite, if they are its place. Further it is clear that the two simultaneous places will be both actually and essentially occupied, not contingently or indirectly.'

> **210a9-13** Further, when water is generated from air, a place will have ceased to be; [for the resulting body is not in the same place; so what kind of cessation is this? The grounds for thinking that place must be something, and also the reasons why one might be puzzled] about what it is, have now been stated.

They say that this argument is capable of raising its own special difficulty about place, because it seems that the place of each sort of body seems to be destroyed, when it turns into another sort. The sign that the place has been destroyed is that the body is no longer

[57] Matter and form, not physical portions. (J.O.U.)

in the same place. But this is absurd. For what would it be for a place to cease to exist? But perhaps the 'further' at the beginning shows that the argument is concerned with the same problem as that immediately previous. If, however, a new problem is being raised, it is better to understand it as interpreted by Aspasius: when water is generated from air, that is, a smaller body from a greater, a part of the place is destroyed. For the place of the water generated is not as great as that of the air from which it came. But it is hard to say in what way place has ceased to be, as we shall learn. This would be of the same kind as the argument from the increase in size of place, where he said 'what are we to say about things that increase in size?'[58] For as he there showed that if there is place, it is necessary that place should increase, here he shows that it is necessary for it to perish. But it is puzzling how place could increase or decrease. For how could the former be true? And it is not a sign that place has perished that a thing is not where it was; for when water is lower down when it has changed from air the place above can remain where it was, even if the same air is not in it.

But it is better to take this same problem as exhibiting the proof that neither matter nor form is place. Alexander says that if we so take it the proof is more concerned with form: 'When changes of bodies into other bodies occur, they lose their form, for water and air have not the same form; but if the form is place, then they lose their place; so they are no longer in the same place. But it is inconceivable what the destruction of place would be. So if the destruction of the form is evident, but the destruction of place is inconceivable, form is 550,1
not identical with place. But if someone says that matter is place when bodies turn into each other, the place is not the same but the matter is. For we say that it remains and receives the different forms.' Alexander thinks that the argument holds better in the case of the form. But, when presenting the hypothesis that form is place, what did he mean when he said, 'So the destruction of the form is evident, but the destruction of place is inconceivable', and thus concluded that place is not form, but in the case of matter he says that the place is not the same but the matter is the same, whereas Aristotle reduces the destruction of place to one absurdity? So perhaps the argument holds equally in both cases regarding the destruction of place when bodies change, whether place be matter or form. For if he did not idly introduce the change of air into water, and, in general, of the greater into the smaller, but did so in order to show that the contraction in magnitude concerned not only the form but also material extension, it is clear that there is the same absurdity in the case of matter and the syllogism would be like this:

[58] Aristotle *Physics* 209a27. (J.O.U.)

if place is matter or form, then in the change of bodies from greater to smaller place is destroyed; but this is impossible, for what would the destruction of place be? He left out the additional premiss as obvious. For into what contrary of place would place disappear, or from what would it come to be? He proves the consequent by saying 'for the body that comes to be is not in the same place'. For if place were supposed to be form, an altogether different place will be found; but if it is supposed to be matter place will decrease in size, and once again will not be in the same place. That the argument proceeds from the partial destruction of place is made clear, I think, by 'for the body that comes to be is not in the same place'. That the destruction of place is reduced to absurdity is shown by 'for what would its destruction be?' This follows also from the common conception of place as not being destroyed with the things in the place, just as vessels are not destroyed with their contents; but while these things remain, different things come to be in them. It also follows from bodies changing their place but places not changing into each other. For the higher place does not change into the lower – how could it? – but higher bodies change into lower bodies. When Eudemus shows that matter is not place he adds this: 'Also matter moves; but place does not move.' It is clear that the same distinction applies to form; for it, too, moves.

CHAPTER 3

551,1 **210a14-24** Next we must determine in how many ways one thing is said to be in another. [In one way it is as the finger is said to be in the hand and generally the part in the whole. In another way it is as the whole is in the parts; for the whole is nothing beyond the parts. In another way it is as man is in animal and generally as the species is in the genus. In another way it is as the genus is in the species and generally as the part of the species is in the complete account of it. Again it is as health is in the hot and cold and in general as the form is in the matter. Again it is as the affairs of the Greeks are in the King's hands and generally in the initiator of change. Again it is as a thing is in its good and in general in its end or what it is for.] But most basically it is as a thing is in a vessel or, generally, in a place.[59]

[59] Many of the uses of 'in' are quite unidiomatic in English translation, even after replacing 'in the King' by 'in the King's hands'. But it is easy to make allowance for

Aristotle has completed the considerations from which it seems necessary that place should be something, and again from which one might be puzzled about its nature and what it is, and whether it exists at all; for if place is not the things that it seems to be one will necessarily puzzle whether it exists at all. So having completed these[60] considerations, and having sorted out our muddled beliefs, he finally prepares for the discovery of the nature of place. But since we come to our idea of place from that of what is in a place, and since being in a place is being in something, he first distinguishes the many senses of 'being in something'. This distinction is particularly useful to him for the solution of Zeno's paradox that place will be in a place, which he said needed a solution. For Zeno said: 'If everything that there is is in a place, it is clear that there will be a place of place and so endlessly.' This proposition will be refuted from the ambiguity of being in something, so that what is in something need not be in a place. So the whole discussion of being in something was completed to this end.

He says that there are eight or nine ways of being in something. Something can be in something (1) as a part in a whole, like a finger in a hand or in the whole body, in the hand as a part and in the whole body as a part of a part, which is also called a bit; (2) as the whole in the parts, like the face in the eyes, nose and the like. For, he says, the whole is not something additional to its parts. For even if the form of the whole is something beyond the form of the parts, so that we speak of the whole and the parts with a formal antithesis, still the substrate is the same; for the whole is not sundered from its parts. (3) Another way of being in something is as the species is in the genus, as we say that man is included in animal, and another (4) is as the genus is in the species, as animal is in man and in horse and the rest of them, since it is included in the account of the species of each. That is why he used 'species' in the singular, not the plural; for being in something as a part in a whole is different from being a species in a genus, as is clear from the fact that the whole is not the whole when one part is removed, but the genus can remain a genus when one species is removed. Also genus is predicated univocally of all the species, but the whole only of things with similar parts, and

this. Thus, if man be defined as a rational animal, then man is 'in' animal, or included among animals, and the generic 'animal' is within the account of man. (The King is the King of Macedon.) Throughout this translation '*meros*' is given its traditional translation as 'part'. But, as elsewhere, this translation is here often inappropriate. Thus it is odd to call form and matter parts of a composite body. This imprecision in English has been judged a lesser evil than giving several different translations, such as 'factor in' and 'feature of', for one Greek word. (J.O.U.)

[60] Reading *toutous* for *toutois*, which appears to be a misprint. (J.O.U.)

then not whole. Also the whole is in all its parts, but the genus in 552,1 even one of its species. That is why Aristotle also in one place said that the whole was in its parts and in another that the genus was in a single species. Further, the genus becomes a part of the account of the species, which shows that the genus is in the species; but the whole is not a part of the account of the part. This is to be expected, since the genus is something common to things already differentiated, belonging equally to all, but the whole is a collection of what remain different elements. (5) The fifth sense of being in something is such as of the form in the matter, as health consists in the hot and the cold. He seems to include everything that is in a substrate in this sense. (6) The sixth is such as the affairs of the ruled being in the hands of the ruler, as the affairs of the Greeks were said to be in the hands of the King. Generally, it is where something is within a prior agency and source of change, as having authority, power and control, as when Homer says

It is in the lap of Zeus.

(7) The seventh sense is as things done for a purpose are in their good and generally in their end, as when all things are in wealth or health which are done for their sake, or as the excellences are in happiness. (8) The eighth sense is as being in a vessel, but since a vessel is said to be a moveable place, he reasonably generalised to being in place as a whole, whether he takes these to be the same sense or two senses.

Alexander says: 'One should note that when he gave health as an example of being in a substrate he added "and generally the form in the matter", the form being in the substrate. However, what is in a substrate is contingently so, but one would say that the form is its essence. Also, what is in a substrate is not a part of the composite thing, as he himself laid down in the *Categories*,[61] saying "what is in something not as being a part cannot exist independently of that in which it is"; but the form is a part of what consists of matter and form.' Eudemus follows him in this; having said 'otherwise affections and conditions will be among existents', he added, 'we must consider whether this is true of figure and generally the shape in the matter', while he himself clearly indicates the difference. So Aristotle seems to take as one the case of form in matter and the central case of being in a substrate, in accordance with some common nature of giving shape. For both give shape to the substrate. Alexander understands the central case of being in as being in a vessel and generally in a place, because 'in something'

[61] Aristotle *Categories* 1a24. (J.O.U.)

signifies 'somewhere' generally, and somewhere is especially a place. So it is central as being common also to the other cases. For the part in the whole is somewhere, and similarly with the other cases. But perhaps, having said how place differs from a vessel, he treats the two cases as having the same force.

These eight or nine significations of 'in something' are stated here. If the commentators enumerate eleven significations in the *Categories*, it should be recalled that there they treat the case of form in matter as one, that of an attribute in a being and generally in a substrate as another, and also being in place as one and in a vessel as another. They thus distinguish ten significations out of thc cases here and add another, being in time, in addition, such as Aristotle being born in the somanyth olympiad. Perhaps he did not add this case now because he had not yet taught anything about time.

A further case of being in something is that of the substrate being in the attribute, as we say that a body is in sickness and a man in a circumstance. It is likely that there are still more significations of 'in something'. But it is clear that 'in something' has many senses and that the expression[62] is ambiguous. Of ambiguous meanings some are on a par as are the meanings of 'I tongue' as 'I kiss' and as 'I utter',[63] for neither sense is more basic than the other; others are not on a par, as when one sense is more basic than another. Thus 'foot' and 'head' are used more basically of animals, less basically and more metaphorically of a mountain.

> **210a25-33** One might raise the problem whether a thing can be in itself, [or whether nothing can, but everything is either nowhere or in something else. This is a double question, depending whether we mean the thing itself or relatively. For when there are parts of a whole, one in which something is, the other that is in it, the whole will be said to be in itself; for it is said to be so relatively to its part. Thus one is said to be white because one's surface is white, and learned because one's intellect is. So a jar will not be in itself, nor the wine; but the jar of wine will be.] For what it is and what it is in are both parts of a whole.[64]

Having enumerated the ways of being in something, he said that it

[62] Expression (*phônê*) reflects the view introduced to the Neoplatonists by Porphyry that Aristotle's account of ambiguity in the *Categories* concerns not things, but words insofar as they signify things.

[63] These examples are substitutes for *erô* in the senses of *philô* (love) and *lexô* (say). (J.O.U.)

[64] Simplicius reads *tou autou moria esti*; the received text of Aristotle omits *esti*. (J.O.U.)

is worth raising the puzzle whether there is still another way such
20 that a thing is in itself. For, if there is, either one way is lacking in the list or it will be subsumed under one of those listed. If there is not, this must be proved. This is specially so because of Anaxagoras' dictum that the cause of the static state of the boundless is that it holds itself firm, i.e. that it is in itself; for nothing else contains it.
25 Aristotle destroyed this view elsewhere by means of the concept of the unlimited. For, he says, if the part is like the whole it should hold firm the part in itself as well, so that everything would be changeless. He also adds other arguments in his same criticism. But here he prefers to discuss being in something, starting his discussion from a thing being in itself in conformity with his project,
30 and showing that it is impossible for a thing to be in itself. As one would expect, the projected proof will also be useful to show that place is not in a place. For if place is in a place *qua* place, it will be in itself. So if that is impossible, place will not be in a place.

In his inquiry whether a thing can be in itself, since a thing might be said to be in itself as such or in respect of something else, he
554,1 agrees that a thing can be in itself in respect of something else. For it can be in respect of a part. For, he says, when there are parts, one being that in which, such as a jar, and the other that which is in it, such as the wine (for these are parts of the combined whole), then it is possible to say that the combination and the whole is in itself,
5 because a portion of it is in a portion. He confirms that a thing is called after its parts from the fact that a body is said to be white because its surface is white, and a man called learned because he has learning in his reason. If, however, the whole be viewed in respect of each of its parts, the whole will be both itself and that in which, such as the jar, and what is in it, such as the wine, and for this reason it will be in itself. So the jar will not be in itself nor the
10 wine in itself, but the jar of wine will be in itself, unless one observes that it is an abuse of terms if those applicable to each part separately are applied to the whole. And if he calls the surface a part of a white body, though it is a limit of it, first one must notice that he also calls the white a part of the white body, although, according to
15 the definition of attributes, it is not a part. For he seems to call anything contributing to the whole a part. Also, if a body is composed of the three dimensions and the surface is one of the dimensions, as Alexander says, why should it not be called a part of the body? But how can the surface be one of the three dimensions, if a surface is something having length and breadth? It is clear that as
20 such it is a part. Even if one takes a body to be something limited, what prevents the limit being a part as contributing to the limited body? It is not, I think, necessary, as Alexander thought, to take 'in itself' as substituted for 'primarily'. The basic antithesis of 'in itself'

is to 'in respect of something else', such as 'in respect to a part' as now, or something external, when the basic is made contingent. In the same way 'primarily' is opposed properly to 'secondarily'. But Aristotle's antithesis is to 'in respect of something else'.

> **210a33-b8** So in that way it is possible for a thing to be in itself, [but it is impossible in the primary sense. For example, white is in the body, since the surface is in the body, and learning is in the soul. It is in relation to these parts, taken as being in the man, that the man, for example, is called white. The jar and the wine separately are not parts, but together they are; so when they are parts the thing is in itself. Thus the white is in the man because it is in his body, and in this because it is in the surface, but it is in this without respect to anything else.] These, the surface and the colour, are different in form and each has a different nature and capacity.

Having shown that in respect of something else it is possible for a thing to be in itself,[65] he will next show that in the primary sense and as such it cannot be. In order that the difference may be clear, he first teaches what *primarily* and *in respect of something else* are. For colour is in a body in respect of something else, but primarily in the surface. And learning is primarily in the intellect, but in the soul 555,1
and the man in respect of something else. So, since the surface is white, and the surface is a part, and the attribution is in respect of a part, the body is called white, though it is not white primarily and as such, but in respect of something else. Also the man is called learned in that way, because when the soul, or the rational part of the soul is called learned, but not as being in the man, this predication is no longer in respect of a part or something else but primary and in respect of itself. That is why he added 'taken as being in the man'. Similarly the jar of wine is said to be in itself in respect of something else, because a part is in a part. But if the wine and the jar are taken in themselves the jar of wine is not said to be in itself. When they are combined, it is. For when they are taken in themselves, they are not parts of a unit and so are not spoken of as such. When they are called parts, since it has been decided to name the whole from its parts, then the jar of wine will be in itself in respect of something else, as the colour is in the body in respect of the surface, and in the surface primarily and not in respect of anything else. However, if the surface is white primarily it does not follow that the surface and the white are the same thing, so that it can be in itself in the basic sense. The accounts of the surface and the white are different, and

[65] Omitting *kai hoti kat' allo adunaton* with aF. (J.O.U.)

so are their natures. So what is primarily in something is not in itself but in something else. This he gathered from the examples given, and it is very important for the proof that nothing is in itself primarily and as such, as we shall learn.

> **210b8-17** From an inductive point of view we see nothing [that is in itself in any of the senses distinguished, and rationally it is clearly impossible. For each thing will have to be both, e.g. the jar both vessel and wine and the wine both wine and jar, if it is possible for a thing to be in itself. So if they were completely in each other, the jar would receive the wine, not as being wine itself but as the wine is wine, and the wine will be in the jar, not as being a jar itself, but as the jar is a jar. Thus it is clear that they are different in essence,] for the accounts of container and contained are different.

He has shown that primarily and as such it is impossible for a thing to be in itself, but in respect of something else and as a part possible, giving the jar of wine as an illustration, and he has drawn a generalised conclusion from the illustrations given that that in which something is and that which is in it are different in kind, and that the surface and the white each have a different nature. He does not wish us merely to adduce other illustrations showing, like the wine in the jar or the white in the surface, that primarily it is impossible for a thing to be in itself. He wishes us also to consider similarly the other listed senses of being in something and investigate inductively in each case whether it is possible in any of the defined senses of being in something for a thing to be in itself. After that we are to investigate by deductive reasoning and collectively whether it is possible for a thing to be primarily in itself, when it has become everywhere clear to us that the container and the contained are different.

For the part in the whole is not in itself, since the part is other than the whole, nor is the whole that is in its parts *qua* parts in itself, since the parts are many and different, but the whole is one. For if the whole is not additional to the parts, still being a whole is one thing, being parts another. Nor can a thing be in itself in the sense of being in a genus; for nothing is its own genus, if genus is different from species. Even when something is in its species as a genus, it is not in itself; for the genus is a part of the account of the species, and it is impossible for a thing to be both part and whole. Nor is the form that is in the matter in itself; for in matter it stands apart from itself. Also, when we say that the affairs of the Greeks are in the hands of the king or of the Lacedaemonians, we say the very opposite, that their government is not in their own hands but of those others. Similarly things that are in their goal are distinguished, since everything before it is dependent on it as being different from it. The case of being in a vessel has been mentioned in advance, for being in a place is still at issue in the question whether a place is in a place.[66] Thus by induction from the ways of being in something we can see that in none of these ways of being in something can a thing be in itself.

One may recognise the same thing by deductive reasoning. For there is a special need of rational demonstration because the senses of being in something were not derived from a logical division, which would make us happy with their treatment *seriatim,* and because it is possible to say that the senses of being in something were taken exclusively from cases of being in something else. Therefore he proves that it is impossible for a thing to be in itself by the second form of hypothetical argument[67] as follows: if there is something that is in itself, then different definitions will belong to the same thing, but this is impossible, therefore there is nothing that is in itself. He also proves the conclusion by 'the jar will be vessel and wine and the wine will be wine and jar, if it is possible for something to be in itself'. For if the jar, the artefact itself, were said to be in itself, then, since what is in the jar is wine, the jar that is in itself will be both jar and wine. Conversely, if the wine is in itself, since that in which the wine is is the jar, the wine will be a jar *qua* container, and wine *qua* what is contained. In this way the same thing will have both the different accounts as container and contained. That this is impossible is obvious; for those very things are different to which different definitions apply. But nothing can be different from itself.

Aristotle used the same illustration very well in the argument where he showed that it is possible for a thing to be in itself in respect of something else, so that it should be clear with regard to the same thing, that it is impossible for a thing to be in itself in such a way that the wine is in itself and the jar in itself and thus the jar of wine in itself. If, as one might claim generally (and this is the force of 'if they were completely in each other'), it is not that each is in the other but that one of them is in the other (for the wine is in the jar), then the jar will receive the wine, not by becoming wine or receiving the account of wine, but by receiving the wine *qua* wine and the wine will be in the jar, not by being the jar itself, but as it is wine and the receptacle is a jar. Each will remain in its own nature and preserve its difference from the other. From this it is clear that they are different in essence and the account of each of that which

[66] Lemma 210b22-31 below.

[67] See n. 6 above.

includes and that which is included is different. From this it obviously follows that a thing is not in itself. For if the including and the included are different, how could a thing be in itself? In this way he showed that primarily and as such nothing is in itself. No one should think that in this way the absurdity results that two definitions fit the same thing, because clearly the jar of wine being in itself is supposed in different ways. If one sense of being in itself is supposed, the jar of wine has one account as being what receives and another as that which the receiver receives.[68]

> **210b18-22** But it is also impossible for a thing to be in itself contingently; [for two things will be simultaneously in the same place. The jar itself will be in itself, if a thing whose nature is to receive can be in itself, and also that which it receives – the wine if that be wine. It is clear] that a thing cannot be in itself in the primary sense.

He has shown that in the primary sense and as such it is impossible for a thing to be in itself, but that it is possible to say this in respect of something else, when a part is in a part and the whole is named after the part. But since what is as such and primarily seems to be the opposite of what is contingently, somebody might suspect that on account of a part, and thus contingently, something might be in itself. Therefore he proves that also contingently a thing is not in itself. For, if we are to speak accurately, 'in respect of a part' and 'contingently' are not the same. Though both are in respect of something else, there is still a difference between them. For a whole
558,1 also may be contingent. For the sailor in the ship is said to move only contingently, but not in respect of a part; for the ship is not a part of him. Conversely, a person is said to see in respect of a part, since he sees with his eyes, but he does not see contingently, since sight belongs essentially to eyes. What is in respect of a part is even opposed in a way to what is contingent. For what is in respect of a part is where a whole is referred to *via* a part, as if a ship were said to creak because its mast did, and as in the case of the jar of wine already discussed; but the contingent is where the part is referred to *via* the whole, as when the sailor is said to move because the ship does and the nose is called snub because Socrates is snub-nosed. For the part and the affection belong to the whole, not the whole to them. So the attribute and the part change contingently, not when they change as such, as when the white becomes black or the sailor walks about in the ship, but when that in which they occur changes. Also iron is said to be hot contingently when the fire in it burns, but

[68] Reading *en hautôi* for *en autôi* in lines 21 and 22. (J.O.U.)

in respect of a part rather than contingently. As I said, being in respect of something else is common to being in respect of a part and to being contingently. So, as in other cases, in the case also of being in something we say that something is contingently in something if it is a happening to or a part of something which is in something as such. For then the happening or the part of that which is within as such is said to be contingently in that within which is that which is so as such. In that way at least the attributes of a body are contingently in a place, such as whiteness and warmth, because the body in which they are is as such in a place. So, thus, if anything were to be contingently in itself, it would have to happen to be a part of that which is in itself as such. From this it follows that if something is contingently in itself, there must first be something in itself as such, which has already been proved impossible. Therefore he has now put aside the adduction of that absurdity.

It is the same with two things being simultaneously in the same place; he refers to bodies, of course, which are the topic under discussion. For if someone were to say that the jar was in itself contingently, since that thing is in something contingently that happens to belong to what is in something as such, then, since the wine is in the jar as such, if the jar is in the wine contingently the jar will be in itself contingently, and the wine, if the jar is a jar of wine, will be in itself as such. That, I believe, is the view of Aspasius. But Alexander writes as follows: 'Thus, if the wine were in the jar, as a contingency or as a part of it, and the jar in the wine as such, in that way the wine would be in itself contingently, which is in no way different from it being so as such.' I do not know what he meant by saying that the wine would be in itself contingently, being aware of Aristotle's statement that the jar would be in itself. It is totally impossible for the wine to be in itself contingently unless it belonged contingently to something present in it as such. In what way would it be more consistent for the wine to be in the jar as such than for the jar to be in the wine, as he says?

The clause 'if that whose nature is to receive can be in itself'[69] created unclarity taken as an incidental remark. He drew attention to another absurdity consequent on the hypothesis, that that whose nature is to receive should itself be in something, rather than something else in it; this is what happens to someone who says that the jar is in itself. It is clear also that if someone were to say that the wine was in itself, the result will be that that whose nature is to be in another will be a container. But perhaps someone will say that the absurdity will arise quite apart from this hypothesis, because it is the jar of wine that is supposed, and it is the jar that is a

[69] Aristotle *Physics* 210b20. (J.O.U.)

container. But this is not the case. Since what is in itself contingently belongs to that which is in itself as such, the latter is a container as such, but what belongs to it is naturally in a container. The remark that 'two things will be simultaneously in the same place' is clearly meant as an impossible absurdity. It seems to be absurd that two bodies should be simultaneously in the same place, and all the commentators that I have come across understand the impossibility as applying to bodies. But perhaps Aristotle omitted to say 'bodies' on purpose and simply meant that the container and the contained would be the two things simultaneously in the same place, being 'different in kind' in that very respect and having 'a different nature', as he said previously.[70] Thus the two definitions of the container and what is in it will apply to the same thing, which is impossible. For those things are different whose definitions are different.

Alexander raises the problem 'what is meant by a thing being in itself neither as such nor contingently. For, if the whole is in the parts and the parts are in the whole, the whole will be in the whole, i.e. in itself. But', he says, 'the conclusion cannot be drawn, for the middle term is not the same. One is the first, when we said that the whole was in the parts, and the other when we said that the parts were in the whole. Also', he says, ' "being in" is to be distinguished as having two different senses, not as generic and specific. So one must distinguish them from each other on the ground that the whole is in the parts throughout, while the parts are in the whole separately. For they are not all together in the whole as one, as the whole is in them.' These are sensible comments. For the predicate of the whole in the minor premiss was 'in its parts', but the parts were predicated of the whole in the major premiss.[71] If then the parts are one thing and being in the parts another, it is clear that, the middle term not being the same, the two extreme terms are not drawn into a
560,1 conclusion. Since this is the nature of the faulty reasoning it would have shown that the genus was in itself, and also the species and the essence and the contingent predicate. For we say that the genus is in the species and the species in the genus, and again that the species is in the individual and the individual in the species, and the contingent in the essence and the essence in the contingent. But in

[70] Aristotle *Physics* 210b6. (J.O.U.)

[71] The syllogism is supposed to be of the form:
Major premiss: M is P
Minor premiss: S is M
Conclusion: S is P
but the middle term M is ambiguous since what we have is:
Major premiss: The parts (M) are in the whole (P).
Minor premiss: The whole (S) is in the parts (M).
and 'in the whole' is not the same as 'the whole'. (J.O.U.)

no case is the middle term the same and therefore the extreme terms cannot be drawn into a conclusion. But why is it reasonable to say that 'the parts are in the whole separately' and not all together? Is it because the conjunction of all the parts is really the whole? For the parts are said to be in the whole when they are considered separately, not when they are scattered everywhere; for thus they would not still be parts.

But Alexander, although he is accustomed to drag in Plato even where it is quite irrelevant, for some reason has here forgotten Plato instead. There is the argument showing that the whole is in itself in the *Parmenides*, running as follows, about the one that has been shown to be the whole:[72] 'If it is thus, will it not both be in itself and in something else? – How so? – I take it that each of the parts is in the whole and none is outside the whole. – That is so. – But are not all the parts contained in the whole? – Yes. – Well now, the totality of the parts of itself is the whole and the whole is nothing more nor less than this totality. – It is not. – And is not the whole the one? – Of course. – So if the totality of the parts is in the whole, and if the totality is the one and so the whole itself, and if the totality is contained in the whole, then the one should be contained in the one, and thus the one itself should at once be in itself.'

Here the argument is that the whole is the totality of its parts, that the totality of the parts is contained by the whole and what is contained by the whole is in the whole, so that the whole is in the whole, i.e. in itself. Also the middle terms used are the same and not different. But since the one has been proved to be the whole, the conclusion is reasonable that the one, as being the whole, is in itself. It is also clear that the other premisses are evident, both that the parts are contained by the whole, and that what is contained by it is in the whole. But the minor premiss that is more crucial to the proof says, 'The whole is the totality of its parts', to which Aristotle himself assents by saying that 'the whole is not something in addition to the parts'. So if both Plato and Aristotle argued correctly, Aristotle that it is impossible for a thing to be in itself primarily, Plato that the whole is in itself, it is clear that Aristotle's proof is concerned with bodies. For if it should be the case that what is properly in itself is not so on account of its parts but as a whole in a whole, but a body is what has separate parts, how should that be a whole in a whole? That is why he took as his examples cases like the jar of wine which clearly has different parts. If it is not a whole in a 561,1 whole, it would not be properly in itself. But if something should be an immaterial whole, which has parts that are not divided up, but distinguished by distinctions that do not require division, each of

[72] Plato, *Parmenides* 145B. (J.O.U.)

which permeates the whole, it is true to say of it that the whole is in the whole and is in itself. In this case it is absolutely true that the whole is the totality of its parts, since the parts remain in the embrace of the whole, so that each, through its undivided unification can be seen as the whole; but in the case of bodies, where each part withdraws into its own perimeter, it is torn away from the whole. So that even if there be not different substrates of the whole and the parts, the whole and the parts are not simply the same thing, because of the separation into parts.

There are two ways of not being in itself, one involving what is inferior to what is in itself, the other what is superior. With regard to the inferior, in so far as in the case of divisible things the different definitions of what receives and what is in it or what contains and what is contained in it do not fit the whole, Aristotle has provided the proof. With regard to the superior, it is because the simplicity of the One does not accept even the inseparable distinction of container and contained, as Plato showed in the first of the hypotheses in the *Parmenides*, where he said:[73] 'But surely that which is in itself, even if it contains itself, is nothing other than itself, if it is to be in itself. For it is impossible for a thing to be in something that does not contain it. – It is indeed impossible. – Would not, then, the container itself be one thing and the contained another? For the same whole will not similarly act and be acted on. In that way the one would be no longer one but two.' It is clear that unless he said this about a different one from that in the previously quoted passage the two would contradict each other. It is implausible that he should have said such manifestly incompatible things about the same things in the first and second hypotheses, in one place denying everything of the one that he affirms in the other.

At any rate, Aristotle showed that body was not the sort of thing to be in itself, and it was about this that he drew his conclusions. It is clear that it is impossible for a thing to be primarily in itself; for neither that which has no double nature, like the One, would be itself in itself, nor would anything having a divided up double nature, like bodies, but only what is both simple and yet with a double nature, since such a thing is in itself, being both simple and double-natured. But how will Plato's proof not fit in with Aristotle's, so that according to him also the whole would be in itself? Perhaps it is because there is a difference in the minor premiss. For Plato says the whole is the totality of its parts because of their inseparable unity; but Aristotle, who says that 'the whole is nothing beyond its parts', sees the identity in the substrate.

[73] Plato *Parmenides* 138A. (J.O.U.)

210b22-7 But Zeno's problem [that if place is something it will be in something is not difficult to solve. For there is nothing to prevent the immediate place being in something, but not as something is in that place but as health resides in warm things as a condition, whereas the heat is in the body as an affection.] So there is no need of an infinite regress. 562,1

Zeno's argument seemed to eliminate the existence of place by raising the question 'If place exists, it will be in something; for everything that exists is in something; but being in something includes being in a place. Therefore a place will be in a place and so on indefinitely. Therefore place does not exist.' So, in embarking on the discovery of place, Aristotle first solves this problem by means of the distinction between ways of being in something that he had sensibly made. He also sensibly proved first that nothing was in itself, in case anyone should think to refute Zeno's argument by saying that body was in a place but place was in itself, and thus the process without limit would come to a halt.

So, instead, Aristotle solves the problem by the distinction between types of being in as follows. If, indeed, being in something was predicated univocally in the case of being in a place, Zeno's problem would have been insoluble, whether it is everything that is or the simple that is in a place, or the corporeal, as all the ancients believed, dreaming about the presence of the forms in matter, as Plato says.[74] But, since 'being in something' is equivocal, there is nothing to prevent the place in which a body is itself being in something, but not in a place; for nothing prevents the one being in one sort of thing and the other in another. 'For health is in warm things, but as a condition, but the warm is in a body as an affection'; but the body is in a place.

A condition differs from an affection in so far as the one is a completion and form, as he reminded us earlier, but the other is an easily lost state and does not affect a thing's essence. For in the *Categories* qualitative affections were said to be like that.[75]

So there is nothing to prevent body being in a place but place in something else, just as the boundary is in that of which it is the boundary, and the surface is in the body. But Alexander rightly raises the question in which of the listed ways of being in something the surface is in the body and in general the boundary in the bounded. Is it as an affection, or condition, or generally being in a substrate, or rather as a part in a whole? For the surface is in a way a part, on the one hand as the boundary of the bounded body, on the other as that part of the body distinguished as having length and

[74] Diels refers to *Timaeus* 52B. (J.O.U.)
[75] Aristotle *Categories* 15a21ff. (J.O.U.)

breadth, the body itself being essentially three-dimensional. Also Aristotle said that place was primarily the immediate container, since the air also as being a body is said to be the place of a thing, when its limit is the immediately containing place.

I think that there are things worth investigating in this context, and first the question how Zeno's argument is refuted by the distinction of ways of being in something, since everything that is is in something else. For even if a body is in something as in a place, the place is also in something as a limit, and that in turn will be in
563,1 something and so on without limit. The second question is how he here refutes the argument of Zeno by the distinction of ways of being in something, when Zeno raises the question as follows: 'If place is something, it is in something.' Previously he had said, 'Zeno's problem requires some solution; for if all that is is somewhere, it is clear that there will be a place of place'. For if he said outright that what is is in a place, how was it still possible to solve the problem by the distinction of ways of being in something, saying, as has now been said, that a body is in a place, but the place is in something else, if it, too, by being one of the things that are, is in a place?[76] Also it must be said that in neither of these contexts does he state Zeno's argument in his own words. Zeno said something intermediate that could be taken to refer to being in a place or to being in something, but Aristotle in criticising the problem referred it to being in a place, but in his solution restricted it to being in something. What then is this intermediate? Perhaps it is 'where'; for this can signify both being in a place and being in something. Aristotle himself recalled this 'where' before setting out his second account when he said, 'If it too is something that is it must be somewhere'. But when he questioned the account he changed the 'where' into 'in a place', saying 'If everything that is is in a place'.

But Eudemus gives the following account of Zeno's view: 'Zeno's problem seems to lead to the same position. For it is reasonable that everything that is should be somewhere; but if place is one of the things that are, where would it be? Presumably in another place, and that in another, and thus continuing.' You see that Eudemus also turned the 'somewhere' into 'in a place' for the effective statement of the problem. How he himself construed the 'somewhere', and that he gave it a general significance become clear where he gives his own refutation of Zeno's argument, writing as follows: 'In reply to Zeno, we shall say that "somewhere" is ambiguous. So if he claimed that things that are are in a place, he was wrong to do so; for neither health nor bravery, nor a thousand other things would be said to be in a place. Place, being like these, is

[76] Reading a question mark for a full stop in 563,8. (J.O.U.)

also not as stated. But if "somewhere" be taken differently place also would be somewhere; for the limit of a body is where the body is, since it is its boundary.'

Alexander, also, explaining that earlier passage that says 'most central of all is as being in a vessel and, generally, in a place',[77] says clearly that 'being in something' has the same force as 'being somewhere'. 'For', he says, ' "being somewhere" is common also to the other senses of "in something"; thus the part that is in the whole is somewhere. But "somewhere" has a special and particular application to things in a place and in a vessel.'

With regard to the first problem, it should be said that an independent refutation of Zeno's argument representing place being in a place as an absurdity is from the distinction between being in something or being somewhere. For place will no longer be in a 564,1
place, which Zeno found absurd. But with regard to the problem that I stated which says that even if it is not in a place, still it is somewhere and in something else and that will also be in something else and so on without limit, Aristotle himself will meet it at the end of the problem about place, attacking the very basis of Zeno's question. For he does not agree that the heavens or the universe are somewhere or in anything whatsoever, so that there would be something else beyond them. We shall hear him saying 'the universe is not somewhere'.[78] For the somewhere is also something and there will have to be something beyond it in which it is contained. There is nothing beyond the totality of the universe, nor is there anything outside everything. Therefore everything is within the heavens. So nothing prevents the parts of the whole being one within the other, but the totality could not be in something else. For there is nothing beyond everything, except as a whole within its parts. That is how the process without limit comes to a halt.

> **210b27-31** It is also clear that since the vessel is no [part of what is in it, for primarily it, i.e. that within which, is different, place must be neither the matter nor the form, but something other. For these, both the matter and the shape, are part of what is within.] Let this be our preliminary statement of the problems.

Zeno's argument sought to prove the regress on the basis of place being in a place. But Aristotle allowed that one thing was in another, but not that place was in a place. Having solved the problem in this way, he saw as a corollary arising from one thing

[77] Aristotle *Physics* 210a24. (J.O.U.)
[78] Aristotle *Physics* 212b14. (J.O.U.)

20 being in another that place was neither matter nor form, as if he had
said: 'Zeno's problem has been solved on the basis of a different way
of something being in something else; but it is clear from one thing
being in another that place is neither form nor matter.' For that
which includes and that which is included are different. This was
said to be the relation of surface to colour and of soul to knowledge.
25 If, then, that which includes and that which is included are different
and if the vessel and place receive that which comes into them
primarily, they are different from it. He uses the vessel rather than
place as being plain to perception and obviously different from what
is in it; and because it is other than what is in it, it is no part of it. If
this is so, place could not be either matter or form, but is something
30 else. For these belong to what is in a place. He sets this out, already
intending to find in advance the essential features of place, since it
is the most important feature of place that it is different from what
is in it.

Having revolved the arguments from all quarters about place, he
adds 'let this be our preliminary statement of the problems', not
because nothing has so far been said by way of demonstration – for
565,1 many things have – but because even the being of place has
appeared problematic because its essence is so hard to grasp.

CHAPTER 4

210b32-211a6 What, then, place is may become clear as follows. [Let us assume about it whatever seems truly to belong to it essentially. We postulate first that place is the container of that of which it is the place, that it is no part of the thing, and that the first container is neither greater nor smaller; also that it is not left deserted by each thing and is separate from it.[79] In addition to this, every place has an up and a down, and each sort of body moves naturally to rest in its own place,] and this it does up or down.

5 Having turned finally to the discovery of place, Aristotle will try to
draw a conclusion on the definition of place from its essential
attributes, since the argument has appeared difficult on all sides, as
is the case when definitions are not immediately apparent. He first

[79] At 211a2-3 the text of Ross reads '*eti apoleipesthai hekastou kai khôriston einai*' which is translated in the Oxford translation as 'place can be left behind by the thing and is separable'. Ross's apparatus gives the reading of Simplicius, which is '*eti mê apoleipeisthai kai khôriston einai*', which I translate as 'it is not left deserted by each thing and is separate from it'. Simplicius explains the first few words as meaning that a place is never vacated to leave a void. (J.O.U.)

presupposes six assumptions from common conceptions about place, which have been made acceptable by what has already been said. The first is that place contains what is in the place; this becomes clear from its likeness to a vessel. However he has already indicated by his examples that the precise place is thought to be that which immediately contains. In general everything that is in something seems to be contained by that in which it is. The second assumption is that the place is no part of what is in the place. This is brought to mind by the arguments in which he showed that place is neither matter nor form. The third assumption is that the immediate place is neither greater nor smaller than what is in it, but the same size. For if it were greater it would not be the place only of the thing, and if less not of the whole of it. The fourth is that 'it is not left deserted by each thing and is separate from it', i.e. that it essentially is never without a body, and yet, while having a body, its existence is separate from that of the body in it. Perhaps he postulated that it is never without a body through relying on his disproof of the void.

This reading is more revealing and more novel, and seems to reconcile two opposites. There is another reading that runs 'also it is deserted by each thing and separated'. This is also clear, for being deserted is the same as being separated. It would mean that place was such that when what was in it moved away it was left behind and separated from it.

The fifth opinion about place is that it admits the differences of above and below. It is not that every place you take is both of these, for it is impossible for contraries to concur in the same thing; but rather he was using the conjunctive particle 'and' instead of the disjunctive 'or', so that he means that every place is either above or below. Homer did exactly the opposite, using 'or' instead of 'and', when he said 'either young or old'[80] instead of 'both young and old'; or perhaps he said that every place was distinguished by being 566,1
above and below rather than that the nature of place, not each place, was such as to admit both.

But why did he recall only those two differences in place, even though he himself has earlier said that 'these are the parts and forms of place, above and below, and the others of the six directions'?[81] Perhaps it is because above and below are the especial differences of place as receptive of natural bodies *qua* natural,[82] whereas the other directions are basically applied to animals and the parts of their bodies which have essentially a front and behind and a right and a left and an impulse to move in those directions.

[80] Homer *Iliad* 14, line 108. (J.O.U.)

[81] Aristotle *Physics* 208b12. (J.O.U.)

[82] Because the four natural elements move either up like fire and air, or down like earth and water.

Sixth and last, he assumes that natural bodies naturally move towards their proper places and remain in them; 'naturally' means 'according to their own nature'. This also is obvious, that a clod of earth moves downwards when it moves, not by some force but according to its own nature, and, having occupied its proper place it remains below. The same applies to fire, upwards. For upwards and downwards are proper differences of natural bodies, for the reason given and because, as is said in *de Generatione et Corruptione*,[83] these are modes mainly of natural bodies that do not possess the motion arising from the impulse of the soul. With these assumptions set out before us, he says, we must investigate with the use of the yardsticks described.

It should be known that both Theophrastus and Eudemus presuppose the immobility of place also in their assumptions. It will become clear why Aristotle left this out, when he explains the difference between place and a vessel.

211a7-11 One must attempt to carry out the investigation [so that the essence will be attained, the problems will be solved, and what seem to be the attributes of place will be so. Further the explanation of our discomfort and the perplexities surrounding it will become clear.] In that way each point will be best demonstrated.

After the presupposition of the attributes that seem according to common conceptions to belong to the subject under investigation, these three things are needed for the indisputable account of its essence: first, the assumptions based on common conceptions must clearly be satisfied by the definition offered; secondly, the problems connected with this satisfaction must be solved by it; thirdly, the explanation of the discomfort caused by the problem must become clear from the definition offered, where such discomfort occurs. For this will be a plain criterion of whether its whole nature has been stated.

567,1 **211a12-23** First, then, one must realise [that place would not have been sought for, if there had not been change of place. It is mainly because they are always changing that we believe they are in place. Of this change, one sort is locomotion, another is increase and decrease. For things change their place also by increase and decrease, and what formerly was here has changed to what is greater or smaller. A thing may be in motion either actually and as such, or contingently. Of the

[83] Diels suggests *de Gen. et Cor.* 333b26ff. (J.O.U.)

contingent one sort is capable of moving as such, such as the parts of the body and the nail in the ship; another cannot move as such but always does so contingently, such as whiteness and knowledge, for these change position because the things] they are in do so.

Having presupposed the common conceptions, and having taught us what the definition of place put forward has to be like, he goes on to remind us, first, what we think things properly in place are like, and then what sort of thing it is. The things properly in place are discovered from things that properly change their place. For the things that properly change their place are those that are properly in a place. We gained our conception of place from the exchange of place by bodies, and we composed our conceptions about place from things that change their place, that it is other than what is in place, that it is the same size as it, that it is separate, and that it contains the differences of above and below, towards which things shift and where they rest. So he has brought together these many conceptions into the one that is the cause of them all, that of change of place. He also indicated that the forms of any sort of change of place were two, locomotion, which is pre-eminently change of place, and increase and decrease. For because of these what was formerly in a certain place, perhaps occupying a span of earth, may shift to occupy a smaller or greater interval of place, in the case of increase clearly a greater one, in the case of decrease a smaller one. It is from these three forms of shifting of position that our conception of place came about. For, in fact, if everything remained at rest, and bodies did not change position nor took over nor relinquished space, what conception of place would we have had?

So, wishing to learn about place from things that change their place, he first distinguishes things that change their place as such and those that do so contingently. By 'as such and actually' he means, for example, that each body is as a whole freed from one place when it changes to another, like my body when I walk and a ship when it sails. Those things that move contingently are things belonging to something that moves as such, either as parts like my hand and the nail in the ship, or as attributes like whiteness and knowledge. There is a difference between these cases. For even if the parts do not move as such as within a whole, since it is necessary that things that move as such and change a place for a place should be altogether free, they still can be freed from the whole and move as such. So when parts move with the whole they actually move contingently, but as such only potentially. But attributes of things that move as such neither can move as such nor have that potentiality, as with whiteness or knowledge. For these things cannot be separated off and move as such, since they are altogether inseparable. But these things always move and change their place when that in which they are moves, but not otherwise, either actually or potentially.

568,1 This being so, one must isolate place from things that change their place as such, not from those that do so contingently, except in so far as those things that potentially can move as such sometimes do so actually, such as the separated parts of a whole that are circumscribed as such; for then they come to be in a place as such. On this topic the division of things that move contingently is useful. For as a thing changes its place, so it is in a place. For things that potentially can exist as such, like parts, can sometimes change their place as such and be in a place as such; but things that move only contingently would never be said to change their place in the basic sense nor to be in a place, but only those things that are separate can touch things that are not continuous with them. It is those things that move as such.

It is self-evident that things that move as such and are in a place as such must be separate. But why[84] is it evident that they must touch something? Is it because, unless it were the universe, a thing's outer parts must necessarily touch something, and if it were the universe it would not exchange one place for another? But that which is touched on all sides is contained by that which is not continuous with it. So that which is primarily in a place is a whole that changes its place as such, is separate, and touches something containing it and of the same size. To anticipate, one can draw the definition of place from things that move as such and are in a place. For the equal limit of what touches them contains things moving locally and that by which moving things are so contained is itself a place. Therefore place is the limit of that which touches and contains things that move locally. Also quantitative change, which is a kind of change of place, is in the same condition, though that which increases or decreases does not change all of its place, but adds or loses some. For this reason, although this is also a change of place, it is nevertheless not called local because the thing does not change its whole place. This, then, is the conception built into our language. Aristotle also says that we think that the universe is in a place because in its motion ever different parts at ever different times change from position to position.

211a23-34 We say that a thing is in the universe as in a place [because it is in the air and the air is in the universe. And it is in the air not as in all of it, but we say it is in the air because of

[84] Accepting Diels's conjecture (*dia ti*) *de* (sc. *dêlon*) *hoti*. (J.O.U.)

> the extremity of the air that contains it. For if the whole of the air were its place the place of each thing would not be the same size as it; but it is thought to be equal in size, and the first place in which it is is so. So when the container is not discrete but continuous with the thing it is not said to be in it as in a place but as a part in a whole. But, if the container is discrete and touching it, the thing is in the immediate extremity of its container, which is neither a part of what is in it, nor greater in extent, but equal.] For the extremities of things that touch are in the same.[85]

He has so far distinguished things which change their place as such, which are also in a place as such, in that they must be bodily substances, whole and free from continuity with each other and 569,1
contained by something else; he continues the discussion in the same way, moving from the loosely called place to that which is strict and primary, in order that he may prove the presupposed conceptions about place that remain. Now we say that something is in the heavens, i.e. in the universe, as being in a place, because we are contained by it, assimilating bodily containment to being in a place. But we are not immediately in the heavens, nor is that basically our place, but only because we are in the air and that is in the heavens. For the same reasons we are in the air, not because we are in all the air, but on account of the extremity of it that contains us immediately. That is why we are also in the air. It is clear that the whole of the air is not our place, if the place must be the same size as what is in it, and if it were not the same size it would not be the place. What is the same size is the extremity of what contains us on our side, in which we primarily are.

Is, then, place this – the extremity of the container that is adjacent to the contained? Or do we need another addition to the account, which he in fact adds? What contains is either continuous with the contained or separate from it, and if it is continuous with it the contained is not said to be in the container as in a place but as a part in a whole. But when it is separate and touching,[86] the contained is also in the immediate extremity of the container as in a place. That is how he shows the difference between things that are in a whole and those that are in a place, just as a little later he shows the difference between things that move in a place and those that move with a whole.

But Alexander thinks that Aristotle proves in this way that the

[85] Here and below I translate *en tôi autôi* in this unidiomatic way to avoid introducing 'place' where *topos* does not occur. (J.O.U.)

[86] Touching (*haptomenon*) is a closer relation than the adjacent (*prosekhes*) of line 14, and it is preserved in the idea of *sunaptein* in the definition of place below at 212a2-7.

continuous parts of the whole that is in a place are not as such in a place, but in the whole that is in a place. However, I think that he speaks clearly about the separation and continuity of the container in relation to what is in it when he writes: 'so when the container is not discrete, but continuous', and, again, 'when it is discrete and touching'. It chimes better with the text and the object in view to understand him as saying that when the container is continuous with the contained, the contained is in the container as a part in a whole; but when it is separate, then the contained is in the extremity of the container that is immediately adjacent as in a place. With this fit the presuppositions about place that it is no part of what is in place, since it is outside it, and is equal to it, which he showed by 'for the extremities of things that touch are in the same',[87] the extremities being the limits of the container and the contained. But if he were to say 'in the same' as meaning 'in a place' he would clearly be speaking contingently. For the surfaces are not in a place as such. But nor are they in the same as in the same substance, for each surface is in its own substance. But the surface is not in the same in any other of the senses of 'in something'.[88] So Alexander writes:[89] 'It is preferable to understand "in the same" as standing for "together". For those things are said to be in the same which have nothing between them. This sense of being in the same seems to have been omitted from the list of meanings of "in something". For those things are in the same that fit together and neither increase the bulk nor have anything between them. For the limit of both becomes one[90] in the case of touching things that coincide, since at that section there is no interval between them. As in the case of the continuous their individuality is lost. For those things are continuous that have no actual limit between them.' In this way Alexander understands 'in the same' in the sense of 'together', saying that this is extra beyond the listed meanings of 'in something'. But Aspasius understands 'in the same' as 'in a place', but in the way that a jar of wine was said to be in itself, because a part is in a part, and the surface of what is in a place is in the place because the surface has become one through having no thickness. But if it is one, how is it in something? Perhaps as in the body of the container.

[87] Aristotle *Physics* 5.3, 226b23-4, 227a10ff, and 6.1, 231a21-3, offer Alexander's equivalent definition of touching, that things touch that have their extremities together (*hama*). This is meant to fall short of the extremities being one, which yields continuity.

[88] I have found no single idiomatic way of translating *en tôi autôi*; so I have translated it literally as 'in the same' in the preceding passage. (J.O.U.)

[89] 'So Alexander writes' is not in the text. But see 570,7. (J.O.U.)

[90] Having their extremities one is given in *Physics* 5.3, 227a10ff, and 6.1, 221a21-3, as the definition of things that are continuous with each other. Alexander's definition here is different.

211a34-211b5 Also, what is continuous with another body does not move in it [but with it, but what is separate moves in it. It makes no difference whether the container moves or not. Again, when it is not separate it is called a part of the whole, like sight in the eye or the hand in the body, but when it is separate it is like water in a cask or wine in a pot. For the hand moves with the body,] but the water in the pot.

This also, I think, shows that what is in a place cannot be continuous with the place, but free. He proves it from motion in a place, if, indeed, what is in a place moves in the place that contains it, but the continuous does not move in its container, but with it. It is in a place when it is separate; for then it moves in its container, both if the container moves, like a vessel, and if it does not, like place proper. But Aspasius took that which was stationary or moving in a place to be like water fluctuating or rotating.

It is clear that it is not the vessel as a whole that is a place, but the hollow surface of the cask or the pot; but Alexander says: 'This also shows that continuous parts are not in a place, if what is in a place moves, but what is in a whole, like a continuous part, does not move in it but with it.' So interpreting the text, he says that the passage following from 'Again, when it is not separate' until 'it is already plain what place is' is the same as what went before and explains it as a duplicate passage, as if he understood what preceded from 'so when it is not separate'[91] as concerned with the difference between what is in a place and what is in a whole, but what follows from 'what is continuous does not move in it' as being concerned with the difference between things moving in a place and the parts that move with the whole. This passage could speak of the same topic succinctly and with examples, first concerning what is in a whole or in a place, what follows concerning things that move with a whole or in a place. For the continuous would be like being in something, 571,1
because it is called a part in a whole, like sight in the eye or the hand in the body, but the discrete are, for example, in a place or in a vessel, even if he added both the examples of vessels, the cask and the pot. But perhaps he added the wine in the pot as an example of being in a place, but the water in the cask as an example of a thing moving in a place. For when drawn up the water moves in the cask, so that sometimes he sets beside each other things in a place and things in a whole, sometimes things moving. This is to distinguish being in a place from being in a whole in both ways. Aspasius also omits completely the latter passage.

[91] Aristotle *Physics* 211a29. (J.O.U.)

211b5-10 It is already clear from the above what place is. [For there are just about four things of which place must necessarily be one; for it is either form or matter or an interval between extremes, or the extremes themselves if there is no sort of interval beyond the magnitude of the body within.]

He has brought the concept of place to that of the limit of the container towards the contained and unconnected with it, because this both contains what is in it and is equal to it and distinct from it, which are the main requirements for place. He now says 'it is already clear from the above what place is', that it is the limit of the container *qua* containing the contained. That this is so, he next tries to prove by a division, using the division through extra terms which runs as follows: it is either *a* or *b* or *c* or *d*; but it is neither *a* nor *b* nor *c*; therefore it is *d*. He also says that there are four possibilities, of which place must be one. For place must be either the form of what is in place, or its matter, or the interval between the extremes, (which is what certain of his predecessors like the school of Democritus, and of his successors like the school of Epicurus and the Stoics and others mentioned by Plato thought that place was), or else the extremes of the container. If, then, it is none of three of these, as he will show, place must be the one that is left. This interval is said by the school of Democritus and Epicurus to be void, so as sometimes to be filled by a body and sometimes to be left empty; but the Platonists and the Stoics say that it is other than the bodies, but always contains a body, so as never to be left empty.[92]

But Aristotle, since he has not exhibited the necessity of the division, added 'just about'; but since unless the division is shown to be exhaustive the proof will seem invalid, we must try to find the necessity. In that way also the 'just about' will be shown to be the sign, not of an indeterminate concept, but of philosophical caution.

Now if that which changes its place is altered in respect of its place, place must be that in relation to which that which changes its place is altered. But that which changes its place is changed either with respect to something that is in it or in respect of something immediately adjacent to it. But what are in the moving object are its matter and its form, and nothing more, and things immediately adjacent are the limit of its container and the interval within that limit, if there is something of that sort, and nothing more. So one may say that place will be one of those four things. If, then, it be shown that it cannot be any of three of them, neither the matter, nor the form, nor the interval, it will be clear that place must be the limit of the container. Why, then, did he not say 'or the extremes if it

[92] 601,24; 618,24 in the Corollary (translated Urmson) ascribe this view to Strato of Lampsacus and recent Platonists. I am not sure why the Stoics replace Strato here.

is none of the three', but 'if there is no interval beyond the bulk of the body within'? Perhaps it was because the hypothesis of the interval seems more plausible than the other two, that say that matter or form is place. Or, more likely, it was because he had anticipated by showing that two of them were impossible and that place was neither matter nor form; so that, if the interval be disproved, then even if nothing be added about the others, it remains that place is the extremities of the container.

> **211b10-19** The shape seems to be the container because it contains; [for the extremes of the container and the contained are in the same place. So both are limits, but not of the same thing, but the form is the limit of the thing, but place is the limit of the containing body. The interval between the extremities seems to be something because the discrete object contained frequently changes while the container remains, as water leaves a vessel, as if the interval were something beyond the body that is displaced. But that is not so;] but some body or other of those that change place and can adjoin the container takes over.

Aristotle describes the reasons why each of the other three was suspected of being place, and further teaches why these suspicions were incorrect by indicating the differences between them. Form comes first. For since a place seems to contain the body in the place, and the form seems to contain that of which it is the form, place seems to be the form. This is especially so because the form that limits the contained and the limit of the container seem to coincide; for when the two are put together they add nothing since both have no thickness. But it is clear that these, the form of that which is in a place and the limit of the container, are different, if, at any rate, the one is the limit of what is in the place, the other the limit of the container. If, then, place is other than what is in the place, the form of what is in the place must be one thing, the place another. But he has earlier shown in many ways that the form is not place, so it now sufficed to show the difference between the limits of the container and the contained.[93]

Next he adds the reason why place is thought to be the interval between the limits of the container.[94] For since we came to our conception of place from the exchange of position by bodies, by which

[93] cf. 535,13-537,4.

[94] Denying the most popular Greek and modern view that place is a three-dimensional interval is essential to maintaining Aristotle's idiosyncratic view that it is the two-dimensional surface of the surrounding container. This is where Aristotle's main attack begins.

different things came to the same place at different times, the interval seems to be something other than the varying body between the limits of the container, such as the concave surface of a pot, so that it is something that could be emptied and receive the bodies
573,1 that replace each other, as a separate thing. For this also seems to be characteristic of place. He adds an explanation of how one is misled. He exposes the argument by showing that there is no such interval; he says the cause of the deception is that when some body leaves the place another always takes its place, and this contains an interval. People transfer the interval within bodies to place, as though it was the interval, so that if another body did not always come in instead, but the receiving vessel collapsed when the former body went out, as we see in the case of skins that collapse when emptied, we would not have thought that there was some interval within. Yet in the case of air we do not fancy in the same way that my body leaves an interval behind it. For the surface containing me flows together as soon as I move and becomes unified in itself. But as things are, because another extended thing comes in, that can coincide with its container but is not unified with it, the interval seems to remain (that is why he added 'things naturally coincident') but not to be unified with the container. For things that become unified, like air with air and water with water, are not as in a place or as in a vessel in that with which they are joined, nor is there still a place distinct from what is in place.

That some body always comes in instead is shown by the stoppers of vessels and the so-called clepsydrae,[95] from which the contained water does not flow unless the air can find some space in them to flow in through. 'But', says Galen, 'let us suppose that when the water is extracted from the pot no other body flows in to take its place; there will still remain the separate interval between the surfaces of the vessel.' But the supposition is irrational; for it begs the question. We are investigating whether there can be a separate interval, and he supposes that there is a separate interval, without proving that there is but rather inventing an imaginary picture for himself. But anyone who says that is supposing the impossible. For as it is impossible for there not to be an interval where there is a body, so it is impossible for there to be no body where there is an interval; for the body contains the interval. So he who supposes that there is no body in the vessel, but that there is an interval in it, supposes that there is and is not a body in it. Let us, if you will, make the same supposition in the case of touching bodies that are not hollow, that when what is touching is removed nothing touches the body. It is plain that he who says this claims that there is a void,

[95] A vessel that holds the liquid through a vacuum, released by lifting one's thumb.

but does not prove it. It is the same in the case of things that are hollow, that are vessels. For there is always something touching them. A difference of shape does not require a change of nature. So we say in this case that if it could be that there was an interval 574,1
separate from a body, there would be a void, but since, as will be proved, there is no void, it is impossible that when something formerly touching something departs nothing else should come to touch its body, and so in the case of the vessel. In general the notion of an independently existing interval comes about when we abstract in imagination the qualities through which one thing is water, another wine and another something else, and when we take the interval of bodies of the same size as something different from each of those things in which it had its being, and to be identical in each case because of the equality, and we take it to be separate from those things with which it existed, one as the interval of wine, another of water. It is abundantly clear that he who supposes that when the body inside is removed nothing is taken in to replace it requires an existent void. If, however, someone were to say that the interval was in the whole universe, but that nothing in it was or became void, but that when a body went out another always took its place, and because of that there was no void and this interval was place, this is a different supposition, needing different disproofs. It seems that Aristotle directs his main attack against this hypothesis.

> **211b19-29** If there were some interval that existed of its own nature and remained [in the same location, there would be an unlimited number of places. For as the water and the air exchange their place all the parts in the whole will do the same as all the water in the vessel. At the same time the place will be altering, so that there will be another place of the place and many places will coincide. But the place of the part is not a different one, in which it moves when the whole vessel changes position, but is the same. It is in the place that they are that the air and the water, or the parts of the water, exchange places, not that in which they come to be, which is a part of the place] that is the place of the whole of the heavens.

Having stated what persuades people to think that place is an interval, and added the explanation of the mistake, he now turns to the disproof of the theory, having given a rather unclear interpretation of the attempted proof. It runs as follows, being a denial of a consequent according to the second indemonstrable method, which fits together as follows: if place has some nature of an interval that remains as such and is not part of a body, three absurdities will follow; (1) there will be an unlimited number of

places, (2) place will be changing its place, so that there will be another place of place, and (3) many places will coincide; but these are impossibilities; so therefore there will be no such interval. The conditional will be seen to be true if we postulate a vessel containing water or air that transfers from place to place in such a manner that what is in it has its own motion, splashing about or rotating. If, then, the interval runs through the whole of the water, he says that 'as the water and the air exchange their places all the parts in the whole will do the same as all the water in the vessel'. For as we thought of place in the case of bodies exchanging their places, when
575,1 where there was water there comes to be air, as was said before, so, as the parts of the water exchange places with each other as they splash about, each portion will be in a part of the interval, i.e. in place as such, if the whole of the interval runs through the whole of the water. Now those who say that place is the limit of the container hold that the whole is in a place as such, but the parts only contingently, since they are not immediately contained by it, but those who say that place is the interval running through then all say that the places of the parts are immediate and as such; for each is as such in the interval. But things that are immediately in a place are discrete, as has been proved earlier. But the places are also separate. If then the parts of what is continuous are separate, as is clearly the case from the fact that they move as such and are in place as such, there must be an unlimited number of parts and of places. But instead of saying that now the parts will no longer be continuous with the whole, but will be in place as such, Aristotle said that all the parts in the whole will do the same as the whole of the water in the vessel. But the whole of the water altered its place and was discrete in a place as such. So the interval that was postulated as place has become responsible for the parts changing their place as such and being in a place and thus for the parts and the places being separated, and thus of the parts also being unlimited in number in actuality and their places also. But since the discussion is now about place, it is the unlimited number of places that he now has instanced as absurd, and not that of the parts. But if places are unlimited in number in actuality as being intervals, so the magnitude in the ladle will also be unlimited. The same can be said of the unlimited parts in the unlimited places. For that which comprises an unlimited number of magnitudes, will be as great and, indeed, of unlimited size. Thus he concludes that if the interval exists there will be an unlimited number of places in the same thing, as in a ladle.

The second absurdity stated also follows, that 'at the same time the place will be changing, so that there will be another place of the place'. For if the vessel containing an interval in itself were to be moved, this interval will be in that interval into which the vessel has moved, and thus there will be a place of place, which is absurd. The same goes for the third, that 'many places will coincide'. For as the vessel is transferred, the interval within its limits will be shifted with it and will occupy another equal interval, so that the things in the vessels will be in the original intervals, which were said to be places, and in those into which they have come. Thus there will be many places coincident, not only because there will be the original one and the new one,[96] but also because those original places that are many because of the separation of the parts (perhaps unlimited like those into which they change) will also be many in the many other places of the same body.

Do the same absurdities follow also for those who say that place is the extremity of the container? By no means. For when the jar is moved neither the water nor its parts are said to come to a different place. For they are in the same place and do not come into another, as if there were an interval within an interval. How, then, if the wine retains the same place as a whole and in its parts, will it move to Athens from Thasos? Perhaps it does not change its place as such, since it stays in the jar, but it is the jar in which the wine is that moves and changes its place, not the wine nor its parts except contingently; but when the parts move they exchange places in the pot. But if place were the interval, bodies would change their place as such together with the vessels and their parts likewise, changing their intervals separately. Things would come to be in one place after another, not contingently, since each of their places is a part of the greater place, and that also of the still greater, and so on to the greatest place and interval, in which are the heavens. For they suppose a permeating interval in which is the whole universe, and within a part of which each body always takes its place. It is a characteristic of this position that it makes a place a part of a place, as an interval is part of an interval. But this is a digression. The point is that those who say that place is an interval will either make the wine move as such, in that respect differing in no way from the jar, or else they must say that it remains in every sense unmoved as it comes from Thasos to Athens. It is not open to them to say that it moves contingently, since they say indifferently that the jar, the wine and its parts are in the interval. So much to clarify Aristotle's words as much as possible.

But one must listen to Alexander who interprets 'there will be many places coinciding' as follows: 'If place were an interval, the things in the vessel would be at once in their original places and

[96] Assuming some such conjecture as Diels's *kai ho eis hon* for the impossible *ei hos* of E. (J.O.U.)

intervals, and also in those into which they have come together with the vessel. If they were taken back they would likewise all be brought back in the intervals in which they were. Thus unlimited places and unlimited intervals would come to belong to each of the things in the vessel, on each transfer of the vessel as the contents acquire an interval equal to itself, i.e. a place.' But how can this be true? For very likely those who say that the interval is place will not even admit that the interval in the pot changes position with the water, which will always be in one part of the whole interval after the other. And even if one conceded this point, it is still not necessary that the interval into which the pot moves should change position; for it comes into one part after another of the interval. I think that it is better to refer the claim that 'there will be many places coinciding' to the many things shifting and the positions they shift into, as has already been said. But Aristotle's 'But the place of the part is not a different one, in which it moves' is directed against the absurdity that says that there will be a place of place and many places coinciding, which followed from the interval which the water took with it being different from that into which it came. So he says that there is not one place after another, but the same, since place is not the interval but the limit of the container, which remains the same. So he says that, when the whole vessel shifts, the place of the part, in which it shifts with the vessel, is not another besides that into which it comes, as those who hold that the interval is place say, but in the place in which they are, i.e. in the limit of the container, in that the air and the water exchange positions with each other, if both are in the pot and move about in it. But if only water is put into the pot, the parts of the water that exchange places with each other move and are in that place in which they also exchange positions with each other, even if the vessel is shifted elsewhere, and not in that place in which, they say, they come, the interval into which they are shifted, which on this theory is a part of the whole interval. In that way the view that the interval is place seems to be demolished.

It is worth considering whether the idea of there being limitless places was based on the hypothesis that things that were immediately and as such in a place must be altogether discrete and circumscribed wholes, but not contained in a whole. Therefore the parts were said to be in a place contingently, and not as such. The explanation of this is, I think, the pre-existing idea of place that we have that makes it the limit of the container, in which the parts are not there as such nor are close to it. Still more, one might say, the whole itself is not in a place as such so as all of it to be touching the place, but only its limit touches it. But if someone supposes that the interval that runs through the whole of the parts is place, then

nothing prevents the parts being in a place also as such and there is no need for them to be discrete. But, just as we say that they are parts, though not discrete, since divided off they are not parts, so, when we say that the parts of a continuous whole are in a place as such, we shall not be compelled to divide them off. They are in a place, just as they are. But they are indivisible, and are divided off either only formally, like the head, shoulders and arms, or only potentially, like homeomeries.[97] So, since place is continuous, and itself has parts, the continuous body will be in it both as a whole and 578,1
in its parts. But there is no need for a place to be in a place or for there to be many coinciding places, nor for place to change position, on such an hypothesis, if the interval running through everything is one, receiving different bodies in different parts of itself at different times. For when the jar of wine that is in an interval which runs through the body of the jar and through the body of the wine shifts, the interval is not carried along with it but it exchanges it for another. For the interval, even if it has a divisible nature, remains unmoved. So thus place will not change position, nor will a place come into a place, nor will there be many places coincident. Let that much suffice as a contribution to the dissolution of the objections mentioned. Perhaps other arguments will add their voices in favour of the interval a little later.

> **211b29-212a2** Matter, also, might seem to be place, [at least if one regards it in something at rest and not divided but continuous. Just as when something is altered there is something that is now white but was formerly black, and now stiff but formerly soft – which is why we say that matter is something – so place seems to exist through a similar act of imagination, except that in the former case what was air is now water, but in the case of place where there was air there is now water. But matter, as was said earlier, is neither separable from the thing nor contains it,] but place does both.

Having proved that form is not place, and the interval likewise, he finally turns to employing the same procedure in the case of matter. As in the other cases, so here he first states the reasons why somebody might think that matter was place, providing now a more persuasive argument than before. At the same time he interprets Plato's aim, because of which he called matter space and place. After that he adduces the considerations because of which he thinks it to be impossible that matter should be place. Wishing to show the persuasiveness of the likening of place to matter he takes the

[97] Homogeneous stuffs, usually tissues.

hypothesis appropriately. For he says that matter might seem to be place when one attends to things that are at rest in a place, and not changing their place. For the things in the matter are at rest in it and do not move. Perhaps Aspasius says better that he who introduces the likeness of matter and place must brush aside the fact that matter is in flux. For place is at rest and is changeless so long as it exists.

Further, matter might seem to be place if one overlooks some of the postulates about place and were to suppose that what was in a place was continuous with the place, since the form in it is continuous with its matter and is not discrete from it. It is clear that this is contrary to the previous presupposition about place. It is because of this hypothesis, he finally says, that we are led away
579,1 from the true account of place to matter. But it is this sort of similarity that brought us to the idea of matter and to that of place. For we conceive of the existence of matter when we see something to be one and the same which once was soft and is now stiff, and now white but formerly black, and we say that it underlies these attributes. Similarly, we came to the notion of place from the same starting points. For it is because where formerly there was air there is now water. But when he seems to add a minor difference he thereby shatters the whole apparent similarity, saying, 'save that because what was water is now air' and what follows. For if the form that comes into the matter shapes the matter and makes it have its own name, but what is in a place does not shape the place, nor is a part of it, nor makes a unity with it, and the place is not named after it, it is clear that there must be a great difference between matter and place. What is common is that, remaining the same, they receive different things, which is why Plato called matter the place and space and receptacle of forms. Would someone then say that place was also the essence of the contingent? Perhaps this would not be far out so far as being a receptacle is concerned, but the difference is that both things in matter and those in a place are beings within beings.

Having said these things, he finally brings arguments against the hypothesis; or rather, he reminds us with a few words of what he had said formerly in distinguishing place from matter. It is this: place is separate from what is in it; but matter is not separate; place contains what is in it; but matter does not contain but is contained by what is in it. But how can he say that matter is not separate, if forms come into it and go away, as bodies do in place? Perhaps, first, because the separate is separated from the separate; all that which is separated has to remain in order that it may be said to be thus separated. But the form does not remain outside the matter, so it is not separated from it. Therefore the matter is also not separated.

Also in very essence form is in matter and a body in a place, while the matter makes a single thing, the composite, together with the form, changing somehow together with the form, especially according to the Peripatetics, who say that the composite is composed of form and matter. In that way they are not separable. But place even then is by itself, having nothing in common with what is in it. In general, we also must not think of the separation of matter with regard to form alone, but also with regard to that which is contained in matter. For matter is not separate from what is in it in the way that a place is separate from what is in it. For it is not separate from the thing of which it is the matter, as the place is separate from the thing of which it is the place.

> **212a2-7** If, then, place is none of the three, [neither form, nor matter, nor some interval always present as something other than the thing that changes position, place must be the one of the four that is left, the limit of the containing body at which it touches the contained body.] I mean by the contained body that which changes by locomotion.

Having shown that place is none of the three things listed, he finally draws together the premisses of the argument and adduces the inductive conclusion that 'place must be the one of the four that is left, the limit of the containing body at which it touches the contained body'. For the containing body may have another limit. For example, the sphere of the moon contains the sublunary things at its concave surface, at which they are joined to it, but it has also another limit, that of its convex surface that is facing the sphere of Mercury.[98] He says, 'I mean by the contained body that which changes by locomotion', so that it is this that is in a place. For place was recognised from change of place. So it is necessary that the heavens, that change by locomotion, be in place, which he himself will shortly deny. So perhaps 'changing by locomotion' means that which naturally moved at times and is stationary at times, whereas the heavens are not changeable, but for ever changing. Alexander says that 'perhaps locomotion and rotation are different, one being in a straight line and totally exchanging one place for another, the other being in a circle and movement in the same place with only the parts changing place. So it is not in a place as a whole, because it does not change place as a whole.' But Aspasius says that 'it is not convertible; for it is not necessary, if everything in place changes by locomotion, that also everything that changes by locomotion should

[98] The heavenly bodies are thought of as rotating round us in concentric transparent spheres, each having an inner and an outer surface.

be in a place'. But he ought to have paid attention to Aristotle's own statement a little later that 'every body that changes, either by locomotion or by increase is, as such, somewhere'.[99] Aristotle then adds, apparently to solve this difficulty, that 'the heavens, as has been said, are not somewhere as a whole, given that no body contains them'. So, since they are not as a whole anywhere, so they do not move as a whole. For they do not exchange place for place, but change only in their parts, in which way they do move. That is why Aristotle added to his previous comments that 'in the direction[100] that they move their parts have a place. For each of the parts is contiguous to the next.'

Having said that contingently the soul and the heavens are in a place he adds: 'for the parts are all in a way in a place; for each contains another in their circle. Therefore the things above move only in a circle and the whole is not anywhere', clearly *qua* whole; for they do not change as a whole. In the way that they do change, they are somewhere; for in the direction that they move their parts have a place, and, being somewhere, they are in motion. 'For', he says, 'each contains another in their circle.' Therefore they move only in a circle. So each is true, both that a body in a place also changes its place and that what changes its place is in a place. Each holds as the other does.

212a7-14 Place seems to be something important and hard [to comprehend, both because matter and shape appear alongside it and because the change of position of the moving body takes place in a stationary container; for it seems to be possible for there to be an interval in between that is other than the moving bulks. Something is contributed by the air that seems to be incorporeal. For place appears to be not only the limits of the vessel] but also that between them as being empty.[101]

Aristotle has previously said[102] that the definition of place should be given in such a form that 'the problems are solved and what seem to be the characteristics of place will characterise it, and also the cause of the discomfort and difficulties will be manifest'. So, having given the definition of place, he immediately tries to exhibit from it the causes of the discomfort. 'For', he says, 'place seems to be something important and hard to comprehend' because in its examination the

[99] Aristotle *Physics* 212b7. (J.O.U.)

[100] At 212b10 the received text of the *Physics* reads *eph' hôi*; Simplicius has *eph' ho*. (J.O.U.)

[101] Simplicius has *hôs on kenon*; the Aristotelian codices omit *on*; Ross has *hôs kenon* (*on*). (J.O.U.)

[102] Aristotle *Physics* 211a8. (J.O.U.)

three things discussed, matter, form and the interval appear alongside it. For if it is impossible for place to be the things that it particularly seems to be, in that way the discovery of place is difficult and causes discomfort. However, if, when we are seeking one thing, others continually appear alongside it, this becomes a sufficient obstacle to the discovery. It has often been said that form appears to be place, because it contains and becomes one with the place when the two surfaces, that of the container and that of the contained, are fitted together. He again added matter to form, because matter seems to be space and a place of forms as different forms at different times come to be within it, in the same way as different bodies are at different times in a place. For where there was once the form of air, there comes that of water. It has also been often said that the exchange of position of bodies has especially given us the concept of place.

His account of the way in which matter and form seem to be place, often given, was sufficient. But he again reminds us of the reasons why the interval seems to be such, since this hypothesis is more persuasive and little examined. For since the exchange of position of bodies takes place in a stationary container, which Alexander takes to mean an immovable one, because place does not move with what is in it, the fact that the contained body that is always in a containing place contains an interval has given some people the idea that there is another interval beyond those in the bodies that move and exchange a place that is equal to them in bulk. But perhaps he did not mean 'immovable' by 'stationary', but as remaining where it was and not exchanging its position; for it has not been demonstrated that place cannot move, nor is that needed for the concept of the interval. But the place needs to remain the same, since when we abstract the different attributes of the bodies exchanging position we imagine the interval as being the same. But perhaps Alexander did not construe 'stationary' as 'immovable' without a point, but because, even if Aristotle indicated that the interval changed, reducing the argument to absurdity when he said[103] 'at the same time the place will be changing', still those who suppose the interval that runs through the whole universe said that it was immovable, receiving at different times in different portions of itself one body after another.

The air, that seems to some people to be incorporeal, seems to contribute to place's seeming to be some interval. For it is because of this that there seems to be an empty space between the limits of the vessel that receives bodies. The things put in do not seem to be merely within the limits of the vessel, but also within the space

[103] Aristotle *Physics* 211b23. (J.O.U.)

between, and even more so, since it permeates throughout. But the limits of what is in the vessel are immediately within its limits. That is why he said that place does not appear to be only the limits of the vessel. But imagination seems to be responsible, since it leaves an empty interval after the abstraction from the body. One might say that this is clearly vacuous; for if imagination were to banish this interval as well, another interval will still be left, and so on without end. 'One must note', says Alexander, 'that he also calls the vessel the limit of the container.' It is indeed clear that when he says 'place is not only the limits of the vessel' Aristotle is not saying that the vessel is the limits of the container but, rather, the limits of the vessel. But one might draw that conclusion from what follows, if the vessel is place and place the limit of the container.

> **212a14-15** Just as the vessel is a transferable place, so place is an immovable vessel.

So far he has treated place and a vessel as being the same, and had given a common definition of place and a vessel; for each is the limit of the container at which it is in contact with the contained. For the vessel is not the body of the baked clay, but the internal surface, and the utensil has the capacity of a vessel while this surface remains continuous, even if the base of the clay be damaged in some way. So now he adds the specific property both of place and a vessel, that place is immovable in its own nature, but the vessel is transportable. So even if he sometimes treats place and a vessel interchangeably, still each retains its specific property, so that place is an immovable vessel and a vessel a transportable place.

Place is reasonably said to be immovable, being the limit of the container. For *qua* surface it is immovable as such, and *qua* place still more so. For how could a place change its place? But someone might say that the vessel is called a vessel primarily in relation to its inner surface. Perhaps it has its capacity which it has in common with place in relation to its surface, but it is a utensil in respect of its body and being transportable as a body that can itself receive a body. Perhaps he did not presuppose that place was immovable in his assumptions about place because he had a need to treat place as a vessel for a time, since he frequently leads us on from the vessel to place. Theophrastus and Eudemus, however, as has been said earlier, presuppose immobility as one of the assumptions about place. Alexander adds that 'even if a vessel is also a place, still the things in the vessel are within the limit of the body that contains them'.

212a16-21 So when the contents of something moving [themselves move and change position, like a ship in a river, they use the container as a vessel rather than as a place. A place needs to be immovable. Therefore it is rather the whole river that is a place, because it is immovable.] So place is that which is the immediate immovable limit of the container.

He has stated the difference between place and a vessel, and, wishing to articulate it more completely and to establish the immobility of place, he distinguishes their contents, teaching us when we speak of them as being in a vessel and when as being in a place. He says that when a moving body is in something moving, so that it moves with that in which it is, like a ship in a river or a piece of wood that does not cleave the water but is contained by the running water and is carried along with it, then it is in that part of the water that contains it as in a vessel rather than as in a place. But what moves while that in which it is remains immobile is in a place. For place is not the sort of thing to move with what is in it, but remains behind, so that it will not itself need a place as it changes its place. But what moves in something moving, like the things in vessels when they move, clearly are moving only contingently. So, having described moving as in a vessel, he indicates motion in a place and place itself with the same example. For as in this example 'the whole river would rather be the place of the ship',[104] because the whole is immobile and retains the same place, flowing from the same source and arriving at the same end. But he added 'rather' because the river would not be the place of the ship in the basic sense; for the whole water of the river and its surface as a whole is not the limit of the container, but a body. However, the whole is rather so in so far as it does not move with the ship and the whole is its place because the ship is carried along in a part of it.

But perhaps the river also is not the place of the ship, but the whole river itself should be in a place, as Alexander says, since the limit of its container does not move together with it, and the ship is in a vessel. 'Perhaps', he says, 'one might call the whole river the place, not of the ship, but of its contents, namely, the water. Also, if we take the river to be not the water but the surface of the earth that contains the water, it will be more suitably said of it that "therefore the whole river is rather place"; for this is basically the place of the water contained and moving within it, being itself immobile. But this is not the primary place of the ship; rather the ship is in the

[104] The trouble with this revision is that it conflicts with the requirement that the ship's place should touch it (see 568,22; 569,17; 580,3). Aristotle's successor Theophrastus traded on this type of problem, as emerges in the *Corollary on Place* (translated Urmson), 604,5-11. See Sorabji, *Matter, Space and Motion*, ch. 11.

water as in a vessel and the water is in a place which is the bank of the river enclosing it, which is what Aristotle called a river and said was the place, not of the ship but of the water.'[105] If this is so, it is clear that he used the word 'river' in different ways, sometimes of the water, sometimes of the surface containing the water.

So, having shown the difference between being in a place and being in a vessel, and having thus indicated the immobility of place, he finally collects the complete definition from the foregoing, that place is the immediate immobile limit of the container. By 'immediate' he indicates that it is adjoining, which he formerly expressed by 'at which it is conjoined with the container'. For what is said to be in a place is in the inner surface of the containing body and also in that within what contains the container. Thus, if a stone is in water, and the water in immobile air, the stone is contained in the surface of the air; but it is not in this immediately and adjoining. Nor is the place conjoined by it to the contained; therefore the air is not basically its place. The limit of the vessel towards the contained is adjoining, but it is moveable and transportable. Therefore it is a vessel and not a place, unless it becomes immobile. This he also confirms by what follows.

> **212a21-9** On this account the middle of the heavens [and the extremity of their rotation facing us are thought by all to be the basic down and up respectively, since the one is always at rest, while the extremity of the rotating heaven always remains unaltered. So, since the light is what is naturally transported upwards, the heavy downwards, the containing limit towards the middle is below, and also the middle itself, and that towards the extremity is above, and also the extremity itself. Hence place seems to be a sort of plane[106]] and like a vessel and container. Also place is together with the thing, for its limits are together with the limited.[107]

He confirms that place is immobile also from the most important differentiae of place; these are, as previously stated, up and down. He seems to me to construct a syllogism something like this: place

[105] Diels fails to close the quote from Alexander. This point seems to be a reasonable place to do so, though not certain. (J.O.U.)

[106] It is essential to understanding what follows to recall Aristotle's very startling idea that a thing's place is the two-dimensional surface of its surroundings, not something three-dimensional that goes right through it.

[107] The lemma omits this final sentence, which is however discussed at 587,28. The next lemma begins at the beginning of the next chapter. In his commentary Simplicius has only *hama gar ta perata*, omitting *tôi peperasmenôi*. It is not clear whether this is merely an abbreviation or an alternative reading to be translated 'for their limits coincide'. (J.O.U.)

has the differentiae of up and down; that which has the differentiae of up and down has the differentiae of middle and extremity; that 585,1
which has middle and extremity as differentiae has unchanging differentiae; that which has unchanging differentiae is unchanging; therefore place is unchanging. It has already been presumed that place has the differentiae of up and down; for the fifth of the presuppositions about place was that every place has an up and down. He proves that the below and the above are the middle and the extremity from the fact that the light body that is carried towards the extremity round about is said to be carried upwards. Thus fire, which is carried towards the surroundings, is said to be carried upwards and heavy things that are carried towards the middle are said to be carried downwards, like the earth. He reminds us that the middle and the extremities are immobile by saying 'the one is always at rest, while the extremity of the rotating heavens always remains unaltered'. For the middle is the true centre of the universe, and the earth about it is always the same, both in its place and in the maintenance of its wholeness and form, even if it changes in its parts and their number. But the extremity of the rotating body facing us which is the concave surface of the sphere of the moon on our side, in its maintenance of its whole and its parts remains numerically always the same. Even if it moves in its parts, still as a whole it does not exchange place for place, but remains always retaining its unchanged place, which is what the argument is especially concerned to show, rather than its unchanging existence.

One must note that he says that the above is the concave surface of the lunar sphere, so that what changes by locomotion is the sublunar. For the specific differences and the essential parts are constitutive of essence. It is also clear from the stronger case. For if the parts were to move, nothing prevents the whole from being immobile, as in the case of parts in circular motion; but if the parts are altogether immobile, so as to move neither as such nor contingently, it is all the more inevitable that the whole be immobile.

So the immobility of place is confirmed from the fact that while we call above and below the differences of place we call those things above and below that are immobile, namely, the middle and the extreme. The place above will be the extremity of the rotation facing us. For things moving upwards, when they have moved as far as it, are contained by or at it, and that is the limit of the containing rotating body at which it contains the contained.

But what is the place below? For the centre is not a place (for it contains nothing, is simple and is the inmost of all things); nor is the earth a place (for the earth is a body, not a limit). Nor is the limit of 586,1
the earth a place; for that is the shape of the earth, not a place. Now

in concluding these investigations he added: 'the containing limit towards the middle is below, and the middle itself, but that towards the extreme is above and the extreme itself.' Here he says which places are below and above and what things are in the places above and below. For the middle itself, which is the earth, is the body in the place below. But the containing limit towards the middle is the place below, which is the limit towards the earth of the body containing the earth, partly water and partly air. There are also areas at which the earth itself has a hollow whose inner limits become the place of other earth, water, fire and air, and of animals and plants. Since all these things compose the place below and not some distinct surface, he himself said vaguely that the containing limit towards the middle is below and the middle itself. For the middle is different at different places, sometimes earth, sometimes some other element or body, but altogether in the middle is where the whole of the earth is. For one and the same thing may be within the limits of several things as its place, as when a stone is in a river, having a part resting in the earth, a part contained by water and a part sticking out of the water and contained by the limit of the air. That, then, is how the thing in the middle place and the place below are.

He says that that towards the extreme is above and the extreme itself also. There, as with below, extreme is that which is above and is in a place, such as the aether,[108] and towards the extreme is the place above, i.e. the limit of the container that inclines towards that which is in the extreme place, such as the concave[109] surface of the moon. This is towards the aether, and the aether is the last of things in place. The middle is that which is in the place below, but, in the case of the above, he did not call the reverse 'that which is towards the extreme', which is verbally like 'that which is towards the middle', the place above, but the thing in the place above; but the extreme that is like the middle is not the thing above but the above itself.

Alexander offers another interpretation of the text, in which he understands the reference to be in each case to place. 'For', he says, 'since not only that which is in the region below is contained by the body towards the middle, which might be the air, but in part also by

[108] The element of which the stars above are made.

[109] It is rather the transparent sphere carrying the moon round the earth which has a concave inner or lower surface. This surface is a place above, because it is the place of the belt of transparent fire where Aristotle believes comets and meteors are found: it is the surface of what surrounds the fire belt. At the same time it is not the ultimate extremity, but only towards the extremity. The extremity itself is the aether which composes the transparent spheres which carry moon, planets, sun and stars. The hollow surface of the moon-sphere is towards this extremity, as constituting its lower margin.

the surface of the body below, which is the earth, in proving of what things the place below is the limit he called the air the limit of that which contains the middle or that which is towards the below and 587,1
which is below, and in the same way he called above the limit of the fire that contains towards the extremity and the extremity itself. But', he says, 'by "the extremity itself" he does not mean the body of which it is the limit but its limit facing us (for the place above is there); he could say that the limit towards the extreme was the limit of that it contained, such as the aether.' In that way it is possible to understand both statements concerning place as being about the same limit, as one could not about the middle; for that required one limit that was towards the middle and another that was of the middle. Porphyry says that the below is double: 'for it is either the limit of the middle, which is its place, or it is the middle itself, of which the extreme is the place. The above is also double; it is either the limit of the extreme, as of the aether, which is its place, or it is the extreme itself, which is the aether.' Also he seems to have said 'the limit of the middle' either to mean 'the containing limit towards the middle' or 'the limit of the earth, whether outwards or inwards, at which it receives and contains'; and[110] he takes it as a place analogous to the limit of the aether; for we call both the earth and its contents below.

Since place is the limit of the container and receptive of what is in it, it consequently also seems to be a plane, that is, a surface, having length and breadth but no depth.[111] That is why he also called it a plane, since the ancients called every surface a plane. For more recent writers said that the plane was a special type of surface and that the spherical, conical and cylindrical were different from it; they said that the plane was a surface that lies evenly with straight lines upon it.

Place is like a vessel. He seems to add 'and that is why it is thought to be a plane' because of those who say that place is form, being the surface of a body. These, in so far as they approached to the limit and the plane, came near to the concept of place; but to the extent that they meant not the limit of the container but of the thing in the place they missed it. What follows, 'also place seems to coincide with the thing, for their limits coincide', seems to have the same effect and to provide some defence for those who thought that place was form; for a place coincides with the thing in the place, because the limits of the thing in the place coincide with place proper. For there is nothing between them, so that they seem to become one because of the union of the planes. It has already been

110 Reading *periekhei kai topon lambanei* for *periekhei, topon lambanei*. (J.O.U.)

111 Simplicius brings out the two-dimensional implications more sharply than Aristotle.

said that 'coincide' does not mean that they are in the same place but that they run together into a unity. For how could place be in a place?[112]

CHAPTER 5

588,1 **212a31-212b3** If a body has another body outside it [that contains it, it is in a place, and not otherwise. Thus, even if water were to have such a lack, its parts will move, since they are contained by each other. The totality will move in one way but not in another. For as a whole it does not change its place all together; but it moves in a circle, for this is the place of its parts. Some things move in a circle but not up and down; others,] those that admit condensation and rarefaction, move up and down.[113]

Having defined place as the immediate and immobile limit of the container, and having confirmed that it is immediate and immobile, he well lays down that on this account not every body is in a place, but 'if a body has another outside it that contains it it is in a place, and otherwise not'. So neither will the fixed sphere[114] be in a place, since there is nothing outside containing it; nor will the whole universe, since there is nothing outside it either. For it[115] would have to be a void, and that will be shown to be non-existent, or a body; the external body will either be limited, so that there will be something outside it also,[116] if every body is in a place, and so on without limit, or it will be limitless, which has been already disproved.

So wishing to convince us that the universe and the heavens are not in a place, he tries to reconcile us to still more paradoxical hypotheses. Even if the universe, he says, be supposed to be water, which specially needs to be in a place or a container since it cannot maintain itself without a container, still such a universe would not be in a place. For a totality has absolutely nothing outside it to contain it, of which the limit towards the contained serves as a

[112] On the discussion from 587,28 to the end see the footnote to the lemma on page 584. (J.O.U.)

[113] This is a puzzling passage. The clue to understanding it is to see that Aristotle treats the water as now being the heavens rotating as a watery mass. (J.O.U.)

[114] The outermost transparent sphere which carries the fixed stars as opposed to the wandering planets. It is not itself fixed. Aristotle is forced to deny it a place by defining a thing's place in terms of outer surroundings which contain it. So it cannot strictly change place when it rotates.

[115] Anything outside.

[116] To limit it, and provide it with a place.

place. But the portions of water would be in a place, if they were separate, as now the separate portions of the universe that touch each other and are contained by one another are in a place.

But, since those things that change by locomotion were said to be in place, he therefore reasons from change of place that the parts are in place but the whole is not. He further proves that the parts change their place but the whole remains immobile. So having started from things that are in place he moved to things that change their place, proving the former from the latter. And he says: 'its parts will move, but the totality will move in one way but not in another', from here on speaking about the heavens or the whole universe. For the same consequences would follow even if everything were water. 'For as a whole it does not change all together', because it does not change its place as a whole, because it is not as a whole in a place, because it is not contained by any body from outside, since it is itself everything or the extremity of everything. The reason why the heavens move in a circle but not up and down, because even things that do not exchange one place for another as wholes can move in circles, is that it exchanges place in its parts. For place of the sort found in things moving in a circle, in which the parts touching one another form places for each other at their limits, is private to such parts, and not to the whole. For nothing outside touches the whole. Therefore the whole cannot move up and down, for the whole would exchange place for place, but moves only in a circle. Of the parts of the whole, those that move in a circle are in the heavens themselves. Other things that admit condensation and rarefaction move up and down, namely the four elements that change into each other.

However, if everything that moves by locomotion was said to be in a place, but what moves in a circle moves by locomotion, how could it not be in place? Alexander says: 'Perhaps "by locomotion" is not the same as "by revolution". But if revolution is in a circle there will be revolution of the parts, not of the whole. For it is the parts that exchange their places.' But, if one were to be precise, not even the parts of the revolving body are in place or change their place, whether as such or contingently. They are not so as such, because they are continuous and not discrete, and they are not so contingently, because those parts would be contingently in place whose wholes were in place as such. But the totality is not in a place, so neither are its continuous parts, either as such or contingently. What, then, does he mean by 'for this is the place of the parts'? Are we to understand these to be the adjacent parts, such as the spheres which are contained one by another? For it is impossible to say this of those that are continuous, since they are not in place.

Again, if the parts become each other's places, how could place be

immobile, when the parts and their limits are in motion? But the parts are to be understood to be not the bodies of the outmost sphere that are continuous with it, but those that are contained by it; for the concave surface of the fixed sphere becomes a place adjoining the sphere of Saturn, and its sphere the next, and it is clear that in so far as they are carried around with the fixed sphere they would be in it as in a vessel. But in so far as each has its own motion they move in it as in something immobile and are as in a place. Alexander says that the statement 'but it moves in a circle; for this is the place of its parts' was made as contributing to this conception. For it is clear that as it moves in a circle it is in contact only with the contained parts. If this is so the parts it contains are in it as a place. This is Alexander's interpretation with regard to the parts of the totality, also of 'they do not move up and down, but in a circle, while others move also up and down'; the first is true of the planetary spheres, the latter of things in becoming. For even if 'but in a circle' was said of the totality, I think that 'in a circle' is more suitably said of the totality of the heavens, which he treats as the same as the totality. Some texts read not 'while others' but 'those things move up and down';[117] if this reading is more satisfactory, then also 'for this is the place of the parts' has to be taken with 'for as a whole it does not
590,1 change its place all together' so that he is saying 'that the totality or the heavens do not change their place all together as a whole; for they do not move up and down, but only in a circle, always retaining the same place; but the parts move both upwards and downwards.

Also he created much unclarity in the argument by using 'the totality' instead of 'the heavens'. For how could things that move up and down be parts of the heavens? But perhaps he speaks of the heavens and the totality as being the same thing in order to have as parts either of the totality or of the heavens the things within it, not itself being in a place. But if the heavens extend right to the centre[118] they really are the totality and everything else is a part of them. Perhaps that is why he later wrote 'nor is anything outside the totality, and therefore everything is in the heavens. For perhaps the heavens are the totality.'[119] For insofar as they contain all things or extend through all things, so far everything is in them, and

[117] Ross's text reads: *kuklôi d' enia: ta de kai anô kai katô* – 'some things move in a circle, others also up and down'. Alexander and Simplicius have: *kuklôi de: enia de kai katô kai anô* – 'but move in a circle; but some things also move down and up'. The alternative reading now mentioned is: *kuklôi de; ta de anô kai katô* – 'but move in a circle; but the parts move up and down' (that admit of condensation and rarefaction). In the main text I depart considerably from Diels's punctuation, but not from his text. (J.O.U.)

[118] On this view the transparent spheres that carry the stars permeate to the centre of the earth.

[119] Aristotle *Physics* 212b16. (J.O.U.)

they are themselves the totality. But insofar as there are other things as well as the heavens, he said that they were 'perhaps the totality'. But the main reason, perhaps, why he calls the heavens the totality is that everything else is insignificant within the totality, since the whole earth is but a central point in relation to the sphere of the fixed stars. That the elements above the earth are not much greater than it is clear from the fact that its shadow extends through everything to the sphere of the moon.[120] That he calls the revolving body as a whole the heavens as one thing is clear from the end of the passage which runs: 'but the aether is in the heavens, but the heavens are in nothing further.'[121]

He seems to his expositors to say both that the parts of the heavens and the heavens themselves are in circular motion, and that the parts are also in a place because they are enclosed, one sphere by another, and have their own private motions, but that the whole is not in a place since it is not contained by anything. It is clear that for the same reason we will not say that the fixed sphere is in a place. For if, under pressure, one were to say that it too, as touching the sphere of Saturn at its concave surface, was in a place, then, first, the container will be in the contained as in a place. Further, using the same argument, one will say that the whole heavens are in a place at the convex surface of the aether; for it touches this too at its own concave surface. In general we shall say in that way that every place is in a place. But if one understands what he there says as being that, while the totality does not change its place because it does not exchange one place for another, it still moves in a circle, but some of its parts move in a circle being within
the fixed sphere, while some, the four elements, move up and down, 591,1
then, first, why are the partial spheres said to change their places although they do not exchange place for place, while the whole is said to move in a circle but not to change its place? Further, he will explain 'this is the place of the parts' in a forced way, though, as has been said, it is better that the totality or the heavens should move in a circle but not change its place, because it does not exchange place for place. But the parts that are sublunary[122] also move up and down; for that is the place where parts change position.

> **212b3-22** As has been said, some things are potentially in place, [others actually. Therefore, when a homogeneous thing is continuous, its parts are in place potentially, but, when they are separated but touch each other like a heap, they are

[120] As shown by lunar eclipse.
[121] Aristotle *Physics* 212b21. (J.O.U.)
[122] Beneath the sphere of the moon are earth, air, fire and water.

actually so. Also some things are in place as such; thus every body that can change whether by locomotion or by increase is somewhere as such, but the heavens, as has been said, are not anywhere as a whole nor in any place, since no body contains them; but in the direction in which they move their parts have a place, for each adjoins another. But some things are in a place contingently, such as the soul and the heavens, since their parts are in a way the totality; for in their circle each contains another. That is why the region above moves in a circle, but the totality is not somewhere. For what is somewhere is itself something, and also there must be something else in addition in which it is and which contains it. But there is nothing in addition to the totality and the whole outside the totality, which is why everything is within the heavens. For perhaps the heavens are the totality. But place is not the heavens but an extremity of the heavens touching the changing body. That is why the earth is in water, that in the air, that in the aether and the aether in the heavens,] but the heavens are finally in nothing else.

Aristotle proposes to prove that the whole heavens and the total universe are not in a place, so he distinguishes all the ways in which something is said to be in a place. For some things are so actually and some potentially. For parts that are continuous and not yet separated[123] are said to be potentially in a place, since, when they are separated they will be actually in a place, whereas what is separate and already cut off, when it is contained by something, is actually in a place.

Again, some things are in place as such, namely those that naturally change their place, either (1) by locomotion, i.e. motion in a straight line, through which a whole changes from one place to another (for locomotion is one thing and revolution[124] another); or (2) such things as change by increase and decrease, which we say are in a place since they, too, change their place;[125] for even if they do not exchange their place as wholes, they take over and relinquish some place. That is why they are in place as such.

Having stated the three senses of 'in a place' – 'potentially', 'actually' and 'as such', he proves that 'the heavens, as has been said, are not somewhere as a whole nor in any place', clearly in any

[123] Diels's text reads: *ta men gar sunekhê kai mêpô merê kata dunamin legetai en topôi*, which must mean 'things that are continuous and not yet parts are said to be potentially in place', which is at best false. I take it that some such word as *diaireta* has fallen out between *mêpô* and *merê*. Aristotle clearly speaks of continuous parts as being potentially in a place in the lemma above. (J.O.U.)

[124] In revolution what revolves need not leave its place.

[125] I reproduce Simplicius' harmless anacoluthon in this sentence. (J.O.U.)

of the three ways of being in a place given. For the heavens are not in a place either in the potential or the actual way, since there is nothing outside which they touch or may potentially touch, as in the hypothetical division. For one cannot even conceive of the totality being potentially, since it is not a part, nor is it in a place as such, as taking over or relinquishing place. Nor does it change its place as a whole, since it does not change either by locomotion or by increase or by decrease. He himself adduced a general explanation of why the whole heavens were not in a place in any of the three ways, that no body contains it. But having said that as a whole they are not in place, because they do not change their place as a whole, he added that 'in the direction[126] in which they move the parts are in a place'. For they move neither up nor down, but in a circle. So the heavens themselves are not in a place, since they do not change their place, but their parts find a place; for the spheres that are for ever within are contained by those outside and move in their own place since they exchange places in their parts. In this way also they differ from the fixed sphere or the whole heavens or the totality, because these do not exchange place even in their continuous parts. For the continuous parts of the fixed sphere, since it is outermost, do not exchange their place, because they are neither in a place nor contained by anything from the outside. But Maximus[127] gave a novel interpretation of 'in the direction that they move their parts are in a place'. 'For the planetary sphere', he says, ' moves to the left, as he shows in Book 2 of *de Caelo*; in this point, then, the parts are in a place, since to the right and to the left are differences of place.'

Having spoken of three of the ways of being in a place, he introduces the fourth. Some things are contingently in a place, such as the soul and the heavens. It is clear that the soul is contingently in place, for the body in which it happens to be as its form is in place as such. But how are the heavens, which are the point of the whole discussion, contingently in a place? Presumably because its parts are in place in a way. They say that he added 'in a way', because of the fixed sphere since that is not in a place. However, if he had made the addition because of that sphere, he would not have gone on to say 'all of them'.[128] For if the fixed sphere is a part of the whole heavens or of the totality, but is not itself in a place, everything cannot be in a place. But perhaps 'in a way' was added to qualify not

126 Reading *eph' ho* as in Diels's text; Ross reads *eph'hôi* in his text of the *Physics*. (J.O.U.)

127 Maximus of Ephesus, Neoplatonist teacher of the Emperor Julian.

128 The order of words in the Greek text of Aristotle is: 'for the parts (are) in a place in a way all of them.' Thus 'in a way' could be taken either with 'in a place' or with 'all of them'. Simplicius first quotes the passage without 'all of them', which implies that 'in a way' applies to 'in a place'. He now tries taking 'in a way' to apply to 'all of them'. (J.O.U.)

'place' but 'all of them', since it is not all of them but almost all of them, if all of them are in addition to the One.

Themistius understood 'in a way' as stated above,[129] because the outermost sphere 'would be in a place at its inside, where it touches and is in a way contained by the sphere of Saturn, but at the outside it totally lacks a place'. It is clear that one who receives this interpretation everywhere will also say that place is in place, because at its inner surface it will touch what is in place and thus the thing in place will be a place. But perhaps, as I said before, neither the totality nor the fixed heavens are in place, even contingently. For, if the totality is contingently in a place, it must be a part or an attribute of that which is as such in place. But the totality is neither a part nor an attribute of anything. The fixed heavens, indeed, are a part of the totality, which is, however, itself not in place as such, so that they could not be in place even contingently. But he seems to take 'in respect of another' as being 'contingently'. For the parts are in a place in respect of another. But 'in respect of another' is common to 'contingently' and to 'in respect of parts'. Since, then, the whole is in its parts, even if not as a part or
593,1 as an attribute, but the parts are in place, then by reference to the parts out of which the whole and the totality consists, the heavens should also be in a place contingently. But perhaps the parts neither of the fixed nor of any other sphere are in a place; for though they adjoin, they are not separated. So if the parts are not in a place as such, then the whole cannot be in a place in relation to its parts.

Alexander says that it is not the fixed sphere nor the divine body alone that is now called the heavens, but the whole universe, since its parts, except those of the fixed sphere, are in place, though the parts of the fixed sphere are not, if 'one contains another in a circle' has to be understood of the parts of the totality, the earth being contained by water, the water in the air, and that in the aether, and that in the heavens, as Aristotle himself says. But perhaps, as Alexander himself noted, what is said will fit even the fixed sphere; for the parts of the fixed sphere also contain each other in a circle. For each of its parts is contained by certain other parts, some preceding in the circle and others following. That is why a thing of this character moves only in a circle and not up and down. 'In the direction in which it moves', as he said before, 'it makes a place for the parts', not being itself a place, but with one part containing another in the circle. So its movement is not locomotion but revolution. Perhaps he set down this 'contain each other in a circle' as suiting both the totality and the fixed sphere and each of the

129 Eliminating the comma between *pôs êkousen* and *hôs eirêtai* in Diels's text. (J.O.U.)

heavenly spheres.

He now shows more clearly that the heavens are not as such in a place, extending their name to cover the totality. For he says that there is the thing in place and outside it that which contains it, as has been said before, but there is nothing outside the totality. For if there is something outside it, it will no longer be the totality, but what includes the thing outside. Moreover, as has been said before, if there is no limitless body, as has been proved, nor a void, as will be proved, what is outside will be a limited body; this will be either nowhere (so that it will be possible for there to be a body that is not in place) or somewhere; and so on without limit, which is impossible both because it is indefinite and because there will again be a limitless body. He seems to say that the heavens are perhaps the totality instead of the universe; for having proved that the totality is not in place, he reasonably added that the universe was the totality. They say that the 'perhaps' is added either because it has not yet been said that there is no void, which some say contains the totality, or because the universe has not yet been shown to be one, or because there are, as such, certain incorporeal forms, or because of philosophical caution. But, perhaps, having shown of the total universe that it is not in place, and wishing to hold that the heavens 594,1
are not in place for the same reasons (for there is nothing outside the whole heavens), he introduced 'the heavens are perhaps the totality' with the addition of 'perhaps' because they contain and because there is nothing outside them and because what is in addition to them is trivial and within them and contained by them, though they are not the totality without qualification. Because everything is in them, they are perhaps the totality.[130]

It should be noted that he now does not call the universe the heavens; otherwise he would not have said 'and that is why everything is in the heavens as in a place'. For everything is not in the universe as in a place but as in a whole. He made it clear that it was in the true heavens that are a revolving body that he said that everything was as in a place, when he said that 'the whole heavens are not a place, but an extremity of the heavens touching things that move in a straight line'.[131] This refers to the moveable body, namely that which is moving freely by locomotion. He called the concave surface of the lunar sphere the extreme, treating all the heavens as one. He makes this clear also by what follows, saying 'that is why the earth is in water', and so on.

At this point Alexander says that one must make it clear that while up till now he has used the word 'heavens' of the whole

[130] cf. 590,5-22. (J.O.U.)

[131] All bodies other than the heavens were held to move (naturally) in a straight line. Cf. 594,25. (J.O.U.)

universe, he now means by the heavens the outermost sphere of the fixed stars. However, even when he said that everything was in the heavens, he was not speaking of the universe, as he made clear by saying immediately that 'it is not the heavens that are place, but an extremity of the heavens'. Yet in all this he does not seem to be speaking about the sphere of the fixed stars, as Alexander says; but he calls the whole divine body the heavens, as is shown by his saying that 'its extreme that touches the moveable body' was place (for he calls moveable those things that naturally move in a straight line) and that 'the aether is in the heavens'. For the aether is not in the fixed sphere, but immediately within the concave surface of the lunar sphere.

Having taken the heavens as one, he reasonably said that the heavens were not in anything else. But Alexander says: 'he called the spheres of the planets the aether; he seems to have omitted fire and placed it with air.' However, he would not have called part of the fifth body aether and part the heavens. For even when he omitted fire he previously said that air was in the heavens: 'since we say that things are in the heavens as in a place, because they are in the air and that is in the heavens.'[132] It is clear that he was calling either the whole universe or the whole of that which revolves the heavens; but he created much unclarity in the passage before us by sometimes saying 'the heavens' and sometimes 'the totality'. Perhaps his discussion refers to both, treating the whole heavens as one thing, and for the same reason – I mean their not being contained by anything – saying equally that the totality and the heavens are not in a place.

Eudemus does not even mention the totality; his enquiry is about the heavens as a whole, writing himself, as I believe, as follows; having said that place must be the containing limit of the container which was immobile he added: 'For what moves is like a vessel, and that is why we refer places to the heavens. For they do not exchange to another place, except in their parts; for they revolve in the same place. But are the heavens themselves in a place or not, and how could either be so? For they are not in place as a whole, unless there is something outside; in that way they would be contained; we shall look into this. But the stars and everything within the outermost body are within its limit at which it contains them. Things thus in something are said to be in place, that in which parts are we call a whole, and it is thus in a place. But a whole is somewhere in another way; for the whole is in its parts, since "somewhere" is ambiguous.'

He has stated in what ways and how contingently the whole heavens are in a place. But it is worth noting what sort of change

[132] Aristotle *Physics* 211a24. (J.O.U.)

revolution is. It is clearly not substantial nor qualitative nor quantitative; so, if it is a change, it remains that it is change of place. But what changes place is definitely in a place, as has often been shown. But Alexander clearly says that this sort of change is not change of place, because no place is exchanged. But he does not say what sort it is, although Aristotle clearly says in *de Caelo* (*On the Heavens*) that it is change of place, which he calls locomotion, which he divides into circular, straight and mixed motion; there he says that 'everything moving locally, does so either in a circle, or in a straight line or a mixture'. Also he clearly says in the eighth book of this work that revolution is change of place.[133]

> **212b22-9** It is clear from the foregoing that all the problems can be solved [on this account of place. For it will not be necessary for place to grow with the thing, nor for there to be a place of a point, nor for two bodies to be in the same place, nor for there to be some corporeal interval (for whatever happens to be within a place is a body, not the interval of a body). Also place is somewhere as well, but not as being in a place, but as the limit is in the limited.] For not everything that is is in a place, but the moveable body.

He has previously said that each thing is best proved if from the
definition given what seem to be the attributes of the thing be seen 596,1
to obtain and if the causes of uneasiness become clear and particularly if the problems be solved; he has also shown that the definition of place as the immediate and immobile limit of the container has this character in regard to the apparent attributes and the cause of uneasiness. He now finally shows how the difficulties raised about place are dissolved.

One of the problems was this: if every body is in place and there is a body in every place, the place of things that increase will have to increase with them; for otherwise the added body will no longer be in a place. It must be said that the containing body may give way, either by exchange when the nearest thing takes over the place of something elsewhere that has shrunk or by its own increase of density and compression; its limits will recede so that the increasing things will get a larger place without place itself increasing. For no limit will, as such, increase. Nor will the body of which it is the limit increase, but on the contrary it may sometimes become compact and more dense.

133 As Diels points out, this quotation does not reproduce accurately anything in *de Caelo*; in that work, at 268b17, Aristotle writes: 'every change that is of place, which we call locomotion, is straight, or circular, or a mixture of these.' It more closely resembles *Physics* 261b28. (J.O.U.)

Also, there will not be the place of a point; for there is no necessity that if there is a place of a body its limits should also be in a place; for they are neither separate nor contained, as a thing in place should be. In general, if not even the continuous parts are in place, as not being separated from the whole, still more their limits will not be in a place. Also it need be no absurdity that a point and a place of a point are different. For if that is absurd, it will also be impossible for bodies to touch each other. For if a point does not coincide with a point, then neither does a line with a line or a plane with a plane.

Thirdly, he says that that difficulty is solved which was first raised, that it is not something incorporeal; for it has three dimensions, length, breadth and depth, by which all body is defined. Also it is impossible for place to be a body, for there would be two bodies in it. This problem also is solved by means of the definition of place. For the limit of the container has not three dimensions,[134] nor is it a body. He gives the same proof that place is not a bodily interval; for a bodily interval has three dimensions, so that for one who says that place is what has three dimensions it will follow that two bodies will be in a place, which does not follow for one who says that it is the limit of a body.

Also Zeno's problem, which he also solved earlier,[135] is solved by the definition. For Zeno said that if everything there is is somewhere, the place will be in a place. But if place is the limit of the container it also will be in something, while 'somewhere' means 'in something'. But it will be so not as in a place but as a limit in the body of which it is the place. So of necessity place will not be in a
597,1 place, if what is in place is a body moving by locomotion. But the limit of a body does not move by locomotion as such, for it is not separated.

212b29-213a11 It is reasonable that each body moves locally into its own place. [For what is adjoining and touching, not by force, is akin to it. Things that are unified do not affect each other, but when touching they affect and are affected by each other. It is not unreasonable that everything naturally remains in its proper place; for so does a part, and what is in place is to its place as a separated part to its whole, as when one moves a portion of water or of air. There is the same relation between air and water, for this is like matter and the former like form, the water the matter of air and the air like an actualisation of water. For water is potentially air and air is in

[134] Place as two-dimensional, again.
[135] Aristotle *Physics* 210b22. (J.O.U.)

a different way potentially water. These things will be distinguished later; they must be mentioned in the present situation, but what is now said obscurely will then be more clear. So if matter and its actualisation are the same thing, for water is both, the latter potentially, the former actually, they will be in a way related as a part to the whole. Therefore these also are in contact, but unification is when both become one in reality.]

That is the end of the account of place, its existence and its nature.

This is not the solution of a difficulty; but the last of the postulates laid down about place was that each body naturally moves locally to and remains in its proper place. So he shows that this, too, follows from the definition of place, while at the same time, I think, resolving a certain suggested objection. That is why he joined it on to the resolutions of the problems. For how, one might say, could the place of the earth be the limit of water, or of water that of air and so on in each case? – at least if a place has to be akin to what is in place and like it – which is why he said 'proper to it', while water is unlike, different from and rather opposite to earth, and air to water, so as even to change into each other. So he says that bodies adjoining and touching each other, not by force, are akin to and related to each other. Air is so related to fire, for they have a common bond of warmth, which is why the change of air into fire is easy. Water is so related to air, for fluidity is common to both; and likewise earth to water, for their kinship lies in being cold; but earth has not this relation to air or water to fire. Even if fire has its being in becoming and in change, still in these ways it is most nearly related to the sphere of the moon and to that portion of it which is closest to the world of becoming. For fire is also alight and immaterial and considered as form compared with the other elements. So each body moves locally towards what is akin as it moves to its proper place.

He well said not only 'adjoining' but 'touching'; for those things adjoin between which there is nothing of the same kind, as he will teach himself,[136] as houses are said to adjoin each other if there is no other house between them. But things that so adjoin do not touch each other nor is one in the other as in a place. So contact is necessary if they are to be in place. It is necessary to add 'not by force'; for things can touch each other not naturally but by force, and these may be neither akin nor touch each other naturally. So if kindred bodies that have a natural order with relation to each other aim to adjoin each other and touch each other, not by force, it is reasonable that when they are torn asunder they should hasten

[136] Aristotle *Physics* 226b34. (J.O.U.)

towards each other as to things related, and remain in each other as related.

598,1 But since the parts are akin to each other and to the whole, but are not in each other or in the whole as in a place, he gives the explanation of this. Things that have coalesced and are continuous with each other can reasonably be said not to act on each other. That which acts and that which is acted on need to be different things, but things that are continuous and coalesced are one. Things that are not unaffected but acting and being acted on are neither continuous nor akin in that way, but do so as touching, the one of them acting, the other being acted upon, close to each other and akin. They are not akin as parts composed of the same elements, but having something in common, something different. That is why they change into each other. So these are the things that are in place. For bodies lying alongside each other, touching, acting and being acted upon are the ones that naturally change and move locally, the less perfect to the more perfect, as he will say.

He has said that the cause of a body reasonably moving locally to its proper place is the kinship of the body the limit of which is its place to that in the place; he also says that the cause of its reasonably remaining in its proper place is the same. For just as what is at a distance desires its kin and moves towards it, so what gets there desires to remain; for it is almost a part of that by which it is naturally contained. He says 'almost' because it is not a continuous part, but discrete. So, as it is natural for parts to move locally towards their wholes and to remain in them, so it is natural for kindred things to move locally towards and remain in their proper place, though discrete. For in that way it is in place. Just as parts in the basic sense, such as a part of water or air, retain within the whole their relationship to it both in motion and in rest, in the same way things akin in the other way retain their relationship, as does water with air. He supposed the part to be moving to improve the parallel with place. For water seems to be a portion of the air, but discrete, and a portion of the composite determined by the form. For air comes to be from water as from matter, as a statue from the bronze, or, rather, as an animal from the seed. For when air comes to be the water does not remain.

But, having said that 'there is the same relation between air and water, for this is like matter and the former like form, the water the matter of air and the air like an actualisation of water', how can he none the less say that water is a part of air? For matter is not a part of form, but both are parts of the composite. Perhaps it is because the composite is mainly given its essence and determined by its form, or, as he himself said, because water is potentially air. For as the seed seems to be a part of the animal that comes from it, so water also is a sort of part of the air that comes from it, being potentially air.

But it was altogether obvious that also water comes to be from air 599,1
and that air is potentially water and is the matter of water, and so also a part of it. But he says that air is potentially water in one way and water potentially air in another. He postpones the precise definition of these to the treatise *On Coming to be and Passing away*. But the transformation of the more material and dense body into the lighter seems to be like coming to be, that of the lighter into the denser like passing away. This is because the elements with lighter parts are analogous to form in relation to those with denser parts. But yet we say that lightness is a merit of water, and denseness of air a defect. Altogether, incompleteness lies with the matter, completion with the form. What is completed can be a sort of part in a whole together with its completion, but what has passed away cannot be a part. So it is reasonable that water aims towards the limit of air, but air, not to that of water, but to that of fire. So, having said that water is potentially air and that water is analogous to matter, but air to form, he finally argues that where the one thing is both potential and actual, and the potential is like to the form and the actual to the matter, that which is actual becomes a part of the potential. So, since water is actually water, but potentially air, and it is transformed into air and in a way exists in it, it stands in a way as part to whole. It would be possible to say the same thing in the case of air, because it is potentially water and is transformed into water, but if this transformation is more like passing away it would be more proper to call it a relation of part to whole in the case likened to coming to be.

He added the reason for including the above when he said 'that is why they touch each other'. For they are in unforced contact between them because of their kinship, and the more material of them is contained by the more form-like as in a place, and this is so when each retains its character in actuality, while being potentially the other. When both become one in actuality, which happens when the one is transformed into the other, as when water becomes air or air becomes water, then[137] they no longer touch each other nor is one in the other as in a place, but they coalesce and become continuous and end up as a whole and parts. When we pointed to their kinship and common qualities to explain their unforced and natural contact, and of one being in the limit of the containing body as in a place, one might rightly conclude, I think, that according to this view water alone ought to be the place of earth and not air. Again, if they tend towards their kin, why does a clod not remain in

[137] Reading *tote* for *touto* as Diels suggests in his apparatus. (J.O.U.)

the air also or in the middle of water, being in its kin and touching it? With regard to the first point, it is possible to say that the air near the earth is turbid and near to being water in its nature, and for that reason it is also akin to earth; with regard to the second, that water also is not akin to the whole of the air, but to the part close to it, and the earth likewise only to the extreme earthlike part of air and water, and that is why water does not remain in the middle of the air nor a clod in the middle of air or water.[138]

But if everything naturally aims at the good and its own welfare, why does it hasten towards that which transforms it and alters it and naturally destroys it? For things which are immediately close to each other change into each other. Perhaps they are not only altered naturally by what is close and transformed into it, but they also alter it and transform it into themselves. So they aim at their own increase. For the truth is not that what is close preys on the other, but that it pursues its own wholeness, and aims at it as being preserved by it. In that way everything whatsoever moves locally towards the place of its whole.

It is worthy of notice that Aristotle left unsolved two problems raised against place, the third and the fourth. The former ran: 'for whatever should we take place to be? For it is not an element nor can it be composed of elements',[139] and the latter: 'further, of what in things should one take place to be the cause' and so on. In reply to the first, one should say that one could raise the same problem about any surface. For it is not an incorporeal point, since it is extended, nor corporeal, since it is not three-dimensional. Also it is not composed of elements for the same reasons; for it is not a body, not being three-dimensional, nor altogether incorporeal since it is extended. So either every surface must be abolished, since these objections are brought against place as a surface, or not place either. For if there is such a nature as to be dimensional, but not in three directions as is a body, what prevents it from being an element in or a part of a body and from being composed of elements which are its extension in two directions, i.e. length and breadth? For if the incorporeal was shown to be the negation of being dimensional, it is absurd and impossible that a surface should be neither body nor incorporeal. But since the one- and two-dimensional is incorporeal,

[138] The statement of the problem at 599,34 seems incomplete and even incorrect. We have an explanation of why water does not remain in the middle of the air or a clod of earth in the middle of air or water. Perhaps the text should read *dia ti mê kai en aeri ho hudôr ê en mesôi hudati menei hê bôlos* – 'why does water not remain in the air or a clod in the middle of water'. We then have 'nor a clod in the middle of air or water' as well as 'water does not remain in the middle of the air' in the explanation because he has explained why earth is akin to some air in reply to the first point. (J.O.U.)

[139] Aristotle *Physics* 209a13ff. (J.O.U.)

it is possible for there to be something that is neither body nor altogether unextended.

In reply to the problem about causes, one should say that if bodies naturally aim at and move locally towards their proper places and remain in them, place must be a perfecting cause as maintaining and completing and looking towards the perfection of each thing. For things are imperfect that are in places foreign to their own wholeness. That is why they move towards their proper places in a straight line when imperfect, as Aristotle himself agrees in *On Coming to be and Passing away*. For no simple thing that is perfect moves in a straight line, but it either remains at rest or moves in a circle.

[Simplicius' commentary on Aristotle's chapters on place is now complete. The whole commentary is interrupted at this point by his *Corollary on Place* in which he offers independent criticisms of Aristotle, states and examines the views of philosophers since Aristotle, and finally offers his own view of the nature of place, which is more different from that of Aristotle than the commentary might lead one to expect. This corollary is followed by a commentary on Aristotle on the void (chapters 6-9), which has not yet been proved not to exist. After this follows the commentary on Aristotle's treatment of time (chapters 10-14), with which this translation continues. The whole discussion of place and time is rounded off by the *Corollary on Time* in which Simplicius criticises Aristotle independently, discusses the views of subsequent philosophers and finally offers his own views about time.[140] – Tr.]

CHAPTER 10

217b29 A discussion of time appropriately follows the foregoing. 694,30

It is obvious that the account of change[141] is of immediate concern to the natural philosopher, given that nature is the principle of change, and that natural phenomena differ from the non-natural by the 695,1
presence of change. But if all change is temporal, and some have thought that time was change itself, others that it was an element in change, it is clear that the account of time is also necessary for the natural philosopher. For time is obvious and on everybody's lips,

[140] The corollaries have been translated in this same series, *Simplicius: Corollaries on Place and Time*, tr. J.O. Urmson, annot. Lucas Siorvanes (London and Ithaca N.Y. 1992).

[141] See n. 1 above.

even if the exact nature of its existence has escaped many. Moreover, like change, it is present in all natural events and in all coming to be. For all these reasons the account of time deserves effort by the natural philosopher.

Aristotle has discussed change in Book 3, adding a treatment of the infinite, since change is continuous and the continuous is divisible *ad infinitum* and either infinite or finite. Then, since all change occurs in place and in time, and since place is more immediate and more obvious than time, he first discussed place, adding an account of the void, since it has been said that the void is place empty of body. Finally, as one would expect, he says that 'a discussion of time appropriately follows the foregoing'.

There is something remarkable about time, in one way in common with other things, in another special to it. For, just as colour and change have a very evident existence, but it is hard to get hold of an account of their nature, similarly the existence of time is totally obvious, not just to the learned but to all. For we all say 'younger' and 'older', and 'yesterday', 'today' and 'tomorrow', and 'formerly', 'at present' and 'in the future'; would one ignorant of days, nights, months and years seem human? But, asked what time is, even the most learned would scarcely have an answer. In this way time is like other things. What is special is that its actual existence is not grasped by perception, even if perception gives us some approaches to it – for the older and the younger, yesterday and today, are not perceptible – and yet it seems the most obvious of all things. In this way the enquiry into its nature has seemed blocked, so that some have been blinded by this and have altogether denied the very existence of time, even though it is so evident. But yet it is out of place to link the abolition of its existence to the ignorance of its nature; for there are many things of which we are ignorant both as to their nature and their existence, but which are none the less among existent things.

> **217b30-2** First it is well to raise the problems about it [quite nontechnically, whether it is an existent or non-existent, and then] what its nature is.

As about other matters, so about time, proposing first an enquiry
into the existence or non-existence of time, he first sets out
nontechnically[142] the arguments that abolish it. Nontechnical, or
696,1 exoteric, arguments are those that are common to all and based on
common opinion, but neither probative nor technical.[143] But he says

[142] For publication outside the school: *exôterika*.
[143] For lectures in the school: *akroamatika*.

that after such an enquiry one must next consider what the nature of time is. For the question of definition follows on that of existence. Alexander says: 'One must point out that on the subject of the void, which he was about to show to be non-existent, he said at the beginning of the discussion of it that he was enquiring into the meaning of the word, but, on the subject of time, which unquestionably exists, into its definition.[144] For non-existent things also, such as the goat-stag, are signified by certain names, but things that have their own nature are fully existent.'

It is worthwhile to examine the merits of this statement. For in the case of time also Aristotle began with the question of existence and not that of definition. Further, even in the case of the unquestionably existent, he raises the question of the meaning of the name before that of existence when the names are not very clear; and we were taught this ordering of the problems in the *On Demonstration*.[145] But since the word 'void' was ambiguous and expressed an unorganised concept, whereas the word 'time' was plain to all, even if the nature of time was not understood, he reasonably started with the meaning of the word in the case of the void, but with existence in the case of time. So we should rather say that it is proper to enquire into the meaning of the name before that of existence, in the case of both the existent and the non-existent, when it is unclear, just as in both cases we should enquire whether they exist or not. But the questions of definition and characteristics and cause arise only in the case of existents. Even in the case of non-existents we do not need to enquire into the meaning of the name if it is clear, as are 'goat-stag' and 'non-existent' itself.

> **217b32-218a8** One might suspect that time either does not exist at all, [or barely and obscurely, for these reasons: part of it is over and no longer exists, part is future and does not exist yet. But both infinite time and any stretch of it consist of these. It might seem impossible for what is composed of non-existents to share in being. Further, if a divisible thing exists it is necessary that all or some of its parts should exist; but of time, which is divisible, part is over and part not yet, and nothing exists. But the 'now' is not a part. For a part is a measure, and the whole must be composed of its parts]; but time does not seem to be composed of 'nows'.

Aristotle sets out three arguments designed to show that time does

[144] A definition does more than state the meaning of a word; it states the nature of an existent thing.

[145] *On Demonstration* is apparently the *Posterior Analytics* and the reference to 71a12ff. (J.O.U.)

not exist. The first of these is as follows: time is composed of the past and the future, of which the former no longer exists and the latter does not yet exist. But what is composed of nonentities is a nonentity. So time is a nonentity. For whether you refer to infinite time or any stretch of it, it consists of these. The second argument requires a presupposition like this: in every compound, if it itself exists, either all or some of its parts must exist; all as in the case of a line, a surface and a body, some in the case of things that have their being in becoming, as in a contest and a change. So, since time, if it exists, is continuous and so divisible, all or some of its parts must necessarily exist. But no part of time does exist, since its parts are
697,1 the past and the future, and neither of these exists, but one is no longer and the other not yet. Therefore time also does not exist. This argument differs from the former, in that that one proceeds categorically, but this one hypothetically, and the first is based on the composition of parts, the second on their division. Accordingly, the first treats time as composite, the second as divisible. What is common to them is that the whole does not exist because of the non-existence of the parts.

But since not only the past and the future seem to be parts of time, but also the present which is the 'now', and that exists, and the foregoing arguments no longer seem to be correct, he properly proves thirdly that the 'now' is not a part of time, introducing two syllogisms in the second figure[146] in a concise form as follows: (1) a part is a measure of the whole; the 'now' is not a measure of the whole of time;[147] therefore the 'now' is not a part of time; it is a property of parts of a whole that the whole is measured by the part: (2) the part is constitutive of the whole, but the 'now' is not a constituent. For time is not composed of 'nows' as the line is not composed of points and, in general, the extended is not composed of the unextended.

Alexander has the reading '*metreitai*' (is measured) in the passive, in the phrase 'for the part is measured', and explains it rather awkwardly, saying that 'it contains an interval; for of what is extended and continuous the parts are also extended; but the "now" is not measurable because it is not extended'. But we should not accept '*metreitai*' in the passive, but rather '*metrei*' (measures, is a measure) as an active verb with the conjunction '*te*' (and). For in the case of a quantity the part is said to be a measure of the whole, but not to be itself measured. Alexander is aware of this reading also.

Aristotle said 'But time does not *seem* to be composed of "nows" '

146 Both syllogisms are in Camestres, of the form 'If all P is M and no S is M, then no S is P'. (J.O.U.)

147 The 'now' is the present instant which, being sizeless, cannot measure or additively constitute time.

because that has not yet been proved, but will be proved in what follows. But since the 'now' is some element in time that exists, but not a part of it, he therefore said at the beginning of the passage that time either 'does not exist at all or barely and obscurely'. For it exists, if at all, in the 'now', which is neither a time, if time is divisible, nor a part of time, if it is not a measure of the whole. One might say 'or barely and obscurely' because some of it is over and some to come, but of that which neither wholly exists nor perfectly exists some cannot be over and some to come.

> **218a8-21** Again, it is hard to see whether the 'now' that seems to divide [the past and the future is always one and the same or ever different. If there are continually different 'nows' and no different parts of time are ever simultaneous, except when one contains and another is contained, as a shorter time by a longer, and if the 'now' that is not, but formerly was, must have ceased to be at some time, then the 'nows' cannot be simultaneous with each other, but the earlier one must always have ceased to be. Now it cannot have ceased to be at itself, since it then was; but it is impossible for the former 'now' to have ceased to be in another 'now'. Take it that it is impossible for the 'nows' to be next to each other, any more than point next to point. So, since it did not cease to be in an adjacent 'now', but in another,] it must co-exist with the infinite number of 'nows' between them; but this is impossible.

Previously he took the 'now' to be an existent, and showed that it is not a part of time. Now, by this third argument he attempts to show that not even the 'now' that seems to exist and to divide the past and the future is anything. For it is necessary, if the 'now' exists, for it to be the same or one after another, which is equivalent to not being the same. If, then, it is not the same nor one after another, it is clear that it does not exist. So first he proves that there is not one after another as follows: if there is one 'now' after another, the earlier must have ceased to be; but the 'now' cannot have ceased to be; therefore 'nows' are not one after another. He proves the conditional from a former 'now' not being simultaneous with a later. For there cannot be two times at once, unless one is longer, the other shorter and one contains the other, as the year is related to the month it contains and the month to the day it contains; for a certain day of the month is said to be present and the month also, and so they seem to be simultaneous. But one 'now' cannot be contained by another, nor can one be a part, the other the whole; for both are without parts and neither longer than the other. Nor are they quantities, but starting points of quantities. But being greater occurs in quantities, so that they cannot be simultaneous, and it is necessary, if they are one after another, for the former to have ceased to be. For what formerly was and later is not has ceased to be.

He then proves the minor premiss that says that it is impossible for the 'now' to have ceased to be, again by means of a division. For, if it has ceased to be, then since what ceases to be does so in time, it must cease to be at itself or at another 'now'. But it cannot cease to be at itself, for then it exists; so, if at another, then, since 'nows' are not adjacent to each other (for time is not composed of 'nows' as he now assumes, but will prove later), but it has ceased to be in another 'now', it is clear that it existed in the interval between itself and that in which it ceased to be. But between 'nows' there is a time, as there is a line between points. So it existed in the time between; but there are infinitely many 'nows' in the time between, given that all time is infinitely divisible, as will be proved, and the division is at 'nows'. So there will be simultaneously the 'now' itself, and those in which it exists, infinite in number, if it existed in the time between. But nor does the 'now' cease to be in the whole of the time in between; for that to be so, a part of the 'now' will have to cease to exist in each part of the time; but it has no parts. So a 'now' does not cease to be either at a 'now' nor in the intervening time. Furthermore, if a 'now' ceases to be at a time other than itself, either that other 'now' also exists and there will be two simultaneous 'nows', that which is ceasing to be and that
699,1 in which it ceases, or yet another. Also it will be either past or future. But, if it is past, the 'now' will have ceased to be before it existed; if it is future, the 'now' would never cease to be but will always be about to cease to be; for the moment of perishing is always future. Time is altogether past and future; but if a 'now' ceased to be in time the simple would be co-extended with the compound. If, however, the earlier 'now' neither exists, since there cannot be two simultaneous 'nows', nor has ceased to exist, it clearly did not exist earlier. But if the earlier 'now' neither was nor is, it is clear that 'nows' cannot be one and another.

218a21-31 But, further, it is not possible for the 'now' to remain always the same; [for there is no finite divisible thing that has a single limit, neither when it is continuous in a single dimension nor when in more. But a 'now' is a limit, and it is possible to select a finite time. Further, if to be simultaneous and neither earlier nor later is to be at one and the same 'now', then, if the earlier and the later are in this 'now', then things that happened ten thousand years ago would be simultaneous with today's happenings, and nothing would be earlier or later than anything else.] Let these be the problems raised about the attributes of time.

Having shown that it is impossible for the 'now' to be one after another, he next proves by two arguments that it also cannot remain always numerically identical. Of these the first is as follows: if there is no finite divisible thing that has a single limit, whether it be unidimensionally continuous like a line, or two-dimensionally like a surface, or three-dimensionally like a body (but rather a line is bounded by two points, a surface by however many lines surround it and a solid by however many surfaces surround it), it is clear that the 'nows' that limit a finite time cannot be one and the same. Rather one is first, the other second. For whatever time we select, even the shortest finite time,[148] will have one 'now' at the beginning and another at the end; for it is bounded at both ends by a 'now'. One must take a finite time to be like a continuous straight line, which is the nature of time. For things like that do not have a single limit. In the case of[149] the sphere and the circle there seems to be a single limit, for they are surrounded by a single surface and a single line.

He shows by a second method that the same 'now' cannot remain for ever. For if there is the same 'now' for ever, everything will be at the same 'now', and neither earlier nor later. Such objects will be simultaneous, and what has so happened has happened simultaneously, and the oldest and most ancient will be simultaneous with recent events and things; for both will be at the same 'now'. What can be more fantastic than this, to be or have come to be now at the time of the Flood? But the long ago and the recent will be of the same date, if things at the same 'now' are at the same date. He quite safely said that the earlier and the later will be occurring at the one 'now'. It is possible for these to have come to be now, but not to exist at a now, for that takes time.

So if it is necessary for the 'now', if it exists, either to be the same or one after another, and if it is neither the same nor one after another, it is clear that the 'now' is not among existents. While drawing this conclusion, he adds that the discussion of the attributes of time, from which he clearly shows that it does not exist, is at an end. In general, if it is necessary that time should be continuous, if it exists, but the 'now' does not join onto either past or future time, since neither of these are actually existing, it is clear that it could not be continuous.

218a31-b9 The definition of time and its nature [are unclear, both in the light of tradition and as regards our foregoing discussion. For some say that it is the motion of the whole,

[148] Simplicius does not believe there is literally a shortest time.
[149] Reading *epi* for *epei* in line 699,24. (J.O.U.)

> others that it is the sphere itself.[150] However, even a part of the revolution is a time, but clearly not a revolution. For what is taken is a part of a revolution, not a revolution. Further, if there were many universes, the motion of any one of them would be a time, so that there would be at once many times. The sphere of the whole seemed so to those who said it was time because everything is in time and also in the sphere of the whole; this view is too simple-minded] to require an examination of its impossibilities.

By the ingenious proofs brought forward by Aristotle that we have discussed, he showed that the existence of time is unclear, since the arguments for its non-existence are so persuasive. He next says that one will find both the definition of time and its nature equally unclear, if one tries to examine them in the light of tradition, as its existence seemed in the discussions which we previously went through.

So its definition is unclear also, given that some say that time is the motion and revolution of the whole, like Plato according to Eudemus, Theophrastus and Alexander. Some say that it is the sphere of the heavens itself, as those Pythagoreans are reputed to say who perhaps misunderstood Archytus who said generally that time was an interval in the nature of the whole, or as some of the Stoics said. Others say that it is change, unqualified. For Aristotle investigates three opinions about time, leaving aside the more mythical. He considers them all together, having distinguished them. For time is either change or that which primarily changes, or the sphere of the whole. If it is change it is either all change or change of the whole; for the concept of time does not admit of changelessness or anything involving it. Of these, he makes a short reply to that saying that time is the motion of the heaven, but a longer to the view that reduces time to being the same thing as change. Aristotle considered the view that time was the sphere itself as too implausible to require a reply.

The reason why they said that the heavens were time was
701,1 apparently that everything was within the heavens and everything was in time. It is clear that those who said this were in the first place ignorant of the ambiguity of 'being in something'. Aristotle himself made the necessary distinctions in his discussion of place.[151] For being in time is one thing, being in a place or in a whole another. Further, they were not expert in logical relationships, which is why he called the view simple-minded. For they argued in the second

[150] The celestial sphere which supposedly carried the stars in a daily revolution around the earth. The whole is the whole spherical universe.

[151] Aristotle *Physics* 210a13ff. (J.O.U.)

figure[152] from two affirmative premisses, that the heaven is that in which everything is, and time that in which everything is, to the conclusion that time is the heaven. It is perhaps possible in the first figure: the heaven is that in which everything is; that in which everything is is time.

Further, the parts of time are the past and the future, but the heavenly sphere has different parts. Also time has its being in becoming,[153] but the heaven does not. Also the parts of the sphere are not everywhere, but time is everywhere. Also the older and the newer are in the same sphere, but not in the same time. Also the sphere is a bodily substance, but time is not. But Aristotle considered this view too simple-minded to require discussion.

But he replied to the other view that said that time was the motion and revolution of the whole. For a part of the circular revolution of the whole, such as a day, is, they say, a time; but it is not the revolution, for a part of the cyclic motion is not a cyclic motion nor a revolution. From this the conclusion is drawn that time is not the revolution of the whole, even if a certain time is a part of the revolution, so that the whole of time is not the revolution. But he showed that a part of a revolution is not a revolution though it is a time, saying that 'it is a part of a revolution, not a revolution'. For a revolution is a motion from and to the same point. The whole is said to have revolved when each of its parts moves from and to the same point. But the time mentioned, such as a day or a night, is a part of a revolution and not a revolution. But, if you take the combination of day and night, this time is greater than the revolution of the whole because of the onward motion of the sun. He adds another point to the argument: for he says that if there were many heavens, i.e. universes, as the school of Democritus supposes, the revolution of each of these would be a time, so that there would be many times at once, which is impossible. It is possible for many motions to be at once, but not many times. For the same 'now' is everywhere the same.

Eudemus sets out further arguments, which it is not unreason- 702,1
able to listen to in his own words, when he says: 'The motion of the whole is clearly in that thing, just as the walking of Diares is in Diares, and thus the motion of the heaven is simply in the heaven. But now we have an absurdity; for time does not seem to be in the heaven but rather the heaven in time. It is also absurd that time should be nowhere else; for the heavens are always in the same place and nowhere else. So there will be no time either on earth or in the sea. But if we say that time is not only the whole revolution, but

152 For the figures of the syllogism, see e.g. the tables in W.D. Ross's commentary on Aristotle's *Prior* and *Posterior Analytics*. Simplicius repeatedly formalises Aristotle's arguments.

153 See Aristotle *Physics* 3.6, 206a22-4.

also that of any one of the parts, while each of the parts is different from the heaven and the heaven from the parts, then the motions of the whole and of each of the parts will be different. But if the motions are different', he says, 'whereas the time is not different but the same, time will not be the motions.'

Also he goes on as follows: 'How are those who say that time is the motion of the sun to speak about the other stars? For they do not all have the same progression; if the progressions were all times, though different, that would be absurd in itself, since there will be many times at once. But if the progressions of the other stars are not also times, one must state the difference, in virtue of which the progression of the sun is time, but not that of the moon nor those of the other stars. This does not seem easy to do.'

Iamblichus, in the eighth book of his commentary on the *Timaeus*, objected to the view as follows: 'If every motion is in time, many motions occur together; but the different parts of time occur separately from each other. Motion occurs round something stationary, but time needs nothing at rest. A motion or rest may be contrary to a motion, either [rest] in the same genus or [motion] in the same species, but nothing is the contrary of time.'

Alexander tries contentiously to show that the view that says that time is the progression of the heaven is Plato's. He first cites Eudemus, who says: 'Plato accepted this view, very surprisingly; for[154] he says that before the heaven came to be there was irregular motion. But "the threads do not join together", given that every change is in time.' He then refuses to accept the view of those who say that Plato, in agreement with Aristotle, holds that time is the number of change, when he calls it 'an eternal image proceeding numerically'.[155] 'For', he says, 'he does not say that time is the number of change, but a numbered, i.e. ordered, change.' Such is Alexander's view.

703,1 But I must indicate, first whence Eudemus got the idea that Plato said that time was the revolution of the heaven, next that it does not involve Plato in the absurdity, as Alexander argued, that there was time before time. 'For', he says, 'if all change is in time, it is clear that discordant and disordered change is in time. So if such change occurred before the heaven came to be, it is clear that there was time before the revolution of the heaven. So, if this is time, there was time before time.'

This requires a reply; for we shall learn what Plato meant when he called time 'an eternal image proceeding numerically', when we have heard Aristotle's statement that time is the number of change.

[154] 'For' offers to explain why it is surprising that Plato should accept, not why he accepted.

[155] Plato *Timaeus* 37D.

Also, I think that he got the idea that Plato held that time was the revolution of the heaven from the following passage in the *Timaeus*:[156] 'So through this the single cyclic revolution became night and day, and a month was when the moon in its circular revolution catches up with the sun again, and a year when the sun completes its circle. But men, except for a few, have not taken notice of the revolutions of the other heavenly bodies, and have not given them names, nor determined their numerical relationships, so that, practically speaking, they do not know that time is the wandering[157] of these bodies.' Thus he seems to say that each revolution is a kind of time, but the single revolution of the universe was the complete time. For he says:[158] 'the perfect number of time completes the perfect year when the relative speeds of all the revolutions come together, measured by the circle of that which moves the same and constantly'. I think that it is clear from these further passages that Plato does not call time the motion of the heavenly bodies, but the measure of motion. For the statement that it was because the revolutions were not numerically measured against each other that their wanderings were not known to be time clearly shows that time has its essence in the measurement of changes. We can learn still more obviously that Plato does not equate change and time when we find him introducing time as the measure of change, when the universe is already in orderly motion through the presence of soul. For he speaks as follows in the *Timaeus*[159] after the animation of the bodily: 'When the father who had begotten it saw it set in motion and alive, come about as an adornment of the everlasting gods, he rejoiced and was glad, and thought to make it like to its pattern. Since that is an everlasting living being, he tried to make this universe as like to it as possible. For the nature of the living creature was eternal. To confer this on the generated completely was not possible. So he thought to make it a changing image of eternity, and, in ordering the heaven, he at the same time made an eternal image, moving numerically, of the unchanging eternal.'[160]

It is clear that this account portrayed the ordered relation of all coming at once to be through the demiurge. For if time came into being together with the heaven, it is clear that neither preceded the other in time. How could anything have preceded time in time? If he added time to a universe already in motion, it is clear that time and

[156] Plato *Timaeus* 39C. (J.O.U.)

[157] *planai*, a reference to the motions of the planets.

[158] Plato *Timaeus* 39D. (J.O.U.)

[159] Plato *Timaeus* 37C-D. (J.O.U.)

[160] Plato in this passage applies the word *aiônios*, translated as 'eternal', both to the timeless paradigm and to the temporal image. In later Neoplatonic writing the universe is called *aïdios*, 'everlasting'. The distinction seems to be sometimes observed by both Plato and Aristotle. See e.g. *Laws* 904A. (J.O.U.)

motion are not the same according to him, but that time is some perfection of motion. Plato says that discordant and disordered motion preceded the birth of the universe, when he says:[161] 'The god wished that all things should be good and nothing should be worthless, so far as possible; so, finding the visible whole not at rest but moving discordantly and disorderly, he reduced it to order from disorder, thinking this a better state than the former.' In saying this he does not mean what he seems to say, that there was once, before time, the disordered visible world, and later, as if waking from sleep, the demiurge put it into order. For, if he constructs through his goodness, and his goodness is always perfect and active and makes all things good by its existence, it is clear that through the goodness of god his constructive foresight existed with it. The passage shows that the god is the cause of harmony and order in the whole, if it postulates it as discordant and disordered before his activity. If Plato supposed in his account a discordant motion before the heaven came to be, and all motion is in time, it is clear that that time was also discordant and disordered, and something like a shadowy outline of time proper is supposed in the story.[162]

But let us turn to what comes next, lest more than is needed be said on this topic.

> **218b9-13** But, since time seems to be a sort of motion (*kinêsis*) and change (*metabolê*),[163] [this view must be considered. Now the change and motion of each thing is in the changing thing itself alone, that is, where the moving and changing thing is itself;] but time seems to be similarly present everywhere and in all things.

He has expounded two theories about time, that which says that time is the revolution of the whole and that which says that it is the
705,1 sphere of the whole, refuting the former and regarding the other more simple-minded one as unworthy of reply. He now sets out the third, the most plausible of the three, which says that time is some form of motion or change. This is clearly a different theory from that which says that time is the revolution of the whole. Alexander now unconvincingly adds: 'This is why Plato and others called time

161 Plato *Timaeus* 30A. (J.O.U.)

162 Such a disorderly time is ascribed to Plato by Atticus and Plutarch with echoes in Galen and Augustine (Sorabji, *Time, Creation and the Continuum*, 270-1). But Simplicius thinks Plato's account of chaos does not correspond to any time, but only to how things would have been if God had not beginninglessly caused order.

163 In this passage 'change' is a translation of *metabolê*, which more obviously than *kinêsis* extends beyond local motion (*kinêsis kata topon*). Indeed, in Simplicius, but not in Aristotle, it extends to change of relation; see *in Phys.* 861,5-28, and Sorabji, *Matter, Space and Motion*, 197-8.

motion. For Plato said, if anything, that time was the revolution of the whole.' For that is how Alexander has so far interpreted him.

But the concepts of time and change seem in a way the same. For time seems to indicate some flow and procession, it has its being in becoming, which is a property of change, and it is an attribute that is similarly continuous and without limit. But Aristotle shows that change and time are not the same, premising that all motion and change are in that which changes alone, or in the place in which that which moves and changes happens to be. Alexander says: 'He has added this because in the case of other changes that with regard to which the change occurs is in that which changes. For what suffers alteration changes in regard to its quality, and what increases with regard to its quantity, and what comes to be with regard to its internal essence. But place, in relation to which that which moves locally changes, is not in that which moves, but where that which changes is, and the motion is there.' So Alexander.

But perhaps local motion is also in that which moves. For it is not the place at which the moving thing moves that is supposed to be time, but the local motion; but motion is always in that which moves, as was proved in the discussion of change. So that is not the reason why 'in the place'[164] is added, but because Aristotle says that not only local motion but also the other types of change are in the place in which the changing thing is. For we say that Socrates was born, grew up and became bald in Athens.

So Aristotle said accurately that all change and motion is only in that which changes or where the changing thing itself happens to be itself. That this is the significance is shown by the other additional premiss which says that time is equally everywhere and present in all things. So if change is not equally everywhere, but only there where the moving thing is, nor present in all things, but only in that which changes, it is clear that time is not change. The formulation is in the second figure. 'Where the moving thing is' is opposed to 'everywhere' and 'in the moving thing only' to 'present in all things'.

218b13-20 Also change is quicker and slower, [but time is not; 706,1
for slow and quick are distinguished by time, the quick being that which changes much in a little time, the slow little in much. But time is not determined by time,] either as a quantity or as a quality of it. So it is manifest that time is not change. [At present it makes no difference whether we say 'change'

[164] At 705,14, where Aristotle's text is expanded, on the basis of Alexander's interpretation. (J.O.U.)

(*kinêsis*) or 'change' (*metabolê*).[165]

He proves the same thing by another argument, again formulated in the second figure, as follows: change is quicker and slower, but time is not quicker and slower; therefore time is not change. That time is not quicker and slower is clear from experience. For we speak of much and little time, but do not speak of it as quicker and slower. But this argument proves it syllogistically: quick and slow are distinguished by time, but time is not distinguished by time; therefore time is not quicker and slower. But it is clear that quick and slow are distinguished by time. For we call that quick which changes much in a little time, slow what changes little in much. But one cannot say that much time has passed in a little time nor little in much.

In fact it is clear that time is not determined by time, also because time is not determined by time as a quantity of itself. For much time is not much by occurring in much time, nor little by occurring in little: that is a property of change. For we say that there is much change when it occurs in much time, and little in little, but not so of time. There is also not a quality of time such as being quick or slow. So its quality cannot be distinguished by time as in the case of change. For we call change quick and slow, and this quality is distinguished by time. For a quick change is that which is much in a little time, and slow that which is little in much. Thus the quantity of change is measured by time, but not that of time. For even if one speaks of much and little time, the distinction is not made by reference to time, but by reference to quantity and number, as will be proved. Plotinus adds another argument to show that time is not change: 'change can cease and be intermittent, but time cannot.'[166]

But, since motion (*kinêsis*) and change (*metabolê*) are not the same thing, as Aristotle will prove at the beginning of Book 5 of this work (for he will prove that coming to be and ceasing to be are change (*metabolê*), but not motion (*kinêsis*)), and since he has used and will use the words 'motion' (*kinêsis*) and 'change' (*metabolê*) indiscriminately, he excuses himself to the reader by making clear that he is doing so.

CHAPTER 11

218b21-33 But neither is there time without change. [For when we ourselves undergo no mental change, or fail to notice the

[165] The final sentence is omitted from the lemma. But Simplicius comments on it as if it were included. (J.O.U.)

[166] Plotinus *Enneads* 3.7.8. (J.O.U.)

> change, no time seems to have passed, as for the mythical people in Sardinia who are said to sleep with the heroes, when they wake up. For one joins the later now to the earlier now and makes them one, cutting out what comes between through unawareness. So, just as there would not be time if the now was not different but one and the same, so also, when the difference is not noticed, the intervening period does not seem to be time. So, if we come to think that time has not passed when we distinguish no change and the soul seems to remain in a single indivisible state, but say that time has then passed when we are conscious and make distinctions,] it is clear that time is not change nor separate from motion and change.

Having shown that time is not change, he next shows that it is not without change, but totally involves change. For he who 707,1
understands points out both the difference and what is common to things naturally connected. That time is not without change, and that one cannot conceive of time without change, he proves as follows: time is imperceptible to us without change; but what cannot be perceived without change cannot be conceived without change, because its existence totally involves change: therefore the existence of time involves change, and without it it cannot even be thought of. That it cannot be perceived without change he proves by saying that when we are mentally totally unchanging, or do not notice ourselves changing mentally, i.e. having some thought, then we think that no time has passed, since one cannot be aware of time without awareness of change. That this is so is clear to us also through sleep; for on waking after undisturbed sleep we are unaware of the intervening time.

This is also clear in intense[167] thought or action; with the mind intent on them, we think that no time has intervened, even though often a long time has passed. For it is not by a change in us but by being aware of change that we are conscious of time being involved in change. This is why, when he said 'when we undergo no change', he inserted 'mental';[168] for the consciousness of time comes about through the mental tracking[169] of the change, which is itself unchanging. One can see it also from the opposite case, since people in pain and distress or in need and want, thinking that such a change is great, think that the time also is great. As the fellow says in the comedy

[167] Reading *suntonous*; Diels has *sun tonous*, a misprint. (J.O.U.)

[168] *dianoia*.

[169] *parakolouthêsis*, a technical Stoic term for understanding, especially introspectively directed. See David Hahm, 'A neglected Stoic argument for human responsibility', *Illinois Classical Studies* 17, 1992.

Will day never come?
For I heard the cock ages ago.

And another says

Those who yearn grow old in a day.[170]

As I said, we ourselves are aware every day from sleep that those not conscious of change are also not conscious of time, but Aristotle confirms it from the longer sleep of those who in the myth sleep in the isle of Sardinia with the heroes. For Hercules was said up to the time of Aristotle and perhaps of Alexander the commentator on Aristotle's works to have had nine sons by the daughters of Thespius the son of Thespieus, who died in Sardinia; their bodies were said to continue whole and uncorrupted, and to give an appearance of being asleep. These are the heroes in Sardinia. It is likely that people slept long sleeps besides them, for the sake of dreams or through some other need, in imitation of them. These people joined the now at the beginning of sleep to that at the end, making them one now, because they were unconscious of change between the two nows; and they joined their notion of time to this now and obliterated the time, because they were unconscious of the change between. He says that just as, if the now at the beginning and that at the end were not different, there would be no time (for the now is not a time but a limit of time), similarly, if we do not notice that the now is different, the time between the two nows seems not to exist.

But Eudemus set aside the confirmation from Sardinia and confirmed the doctrine from an event at Athens. 'For', he says, 'it is stated that at a public religious celebration some people were feasting in a deep cave; they got drunk and went to sleep towards daybreak, together with their servants, and slept through the rest of the night, the next day and also the next night. One of them woke up, saw the stars were out and went back again to rest. On the following day, when they woke up, they celebrated the Koureotis a day later than the other people, which is how they discovered what had happened.' The Koureotis was the third day of the Apatouria, as the first was Dorpia, the second Anarrusis and the fourth Epibda. So after the celebration of the second day they slept through the next and the night after, and observed the fourth day as the Koureotis instead of the third.

So, if it is when we believe there is no change but are at the same now that we think that no time elapses, but it is when we are conscious of change and distinguish several nows, one as beginning,

[170] Aristophanes *Clouds* lines 3-4 and Theocritus 12.3. (J.O.U.)

one as ending, one as earlier, the other later, that we are conscious of intervening time, it is clear that there is no time without motion or change. On the basis of these statements the remarkable Galen, in the eighth book of his *On Demonstration*, suspects that Aristotle says that there is no time without change because we conceive it as changing.[171] He also draws an odd conclusion that 'the altogether unchanging also must involve change, if the thought of these also involves change for us. For we think of nothing with unchanging thought.' One might as well say that there could not be unextended things without extension, if we commonly think of them extendedly, using imagination. But first he ought to have observed that Aristotle does not mean that the soul changes, but that it is active, thinking it proper to call only physical changes changes. So he does not say that time is not without change because we think of it while undergoing change, but because the unchanging thought of change carries with it the thought of time. That this is what he is saying he makes clear also from his statement that even if we change our thought, i.e. even if in thought we are active about change, but 'fail to notice the change, no time seems to have passed'. And, again, he says that we fail to notice the change not because we are not changing, but because we have no awareness of change; and, if he says that we are simultaneously sensible of change and time, he does not say that we change as we are so sensible. And even if someone might think that 'when we ourselves undergo no mental change' means the same as 'when we are not changing mentally', even so he does not think that time involves change because we ourselves change as we are aware of time, but because we track the change in thought, whether or not we ourselves change as we track it. That is why he added the correction 'or fail to notice the change'.

However, as I said earlier, when we work or think intently we clearly undergo change, but we do not pay attention to the extent of the change nor distinguish the earlier and later nows, and thus become ignorant of the intervening time, because it co-exists with the change that we compressed. Yet we cannot think of the poles or

171 According to Themistius' report (*in Phys.* 144,24ff.) of Galen, we conceive time while *we* are changing in our minds (*kinoumenoi* instead of *kinoumenon*). The Galenic interpretation that emerges from the two reports combined is that, according to Aristotle, our minds change when we think, and that is why we have to think of time as changing. Galen's objection to Aristotle, as thus interpreted, is that even something unchanging like the centre of the earth (the example in Themistius) would have, on that showing, to be thought of as changing. You might as well say that if we have to think of something unextended (a point, for example) by using an extended mental image, we have to think of the point as extended. I think it is easiest to take the remark about unextended things as part of Galen's objection. On the punctuation of Diels, the editor, it is part of Simplicius' reply to Galen, but in that case Simplicius will have misunderstood. The translator has kept to Diels's text, so the reader may judge for himself which is the right interpretation.

the centre of the universe without the thought of change. For, in general, in the case of things co-existing and essentially related to each other, awareness of one inevitably accompanies awareness of the other. Thus he who thinks of a friend will realise that he is a friend of a friend, and he who thinks of the right will immediately refer the left to it. So I think one should consider what Alexander meant when he said that in no other case did the thought of one thing accompany that of another, except in the case of time, if the thought of time is conjoined with that of change. But it is clear that change is also carried along with time. For, having travelled a certain time, we judge that in that time we have done so many miles.[172] Aristotle also says in what follows that 'indeed, when a certain time seems to have passed, simultaneously a certain change seems to have taken place'.

> **219a2-10** Starting from this, we must, since we are trying to discover the nature of time, [determine how it is related to change. For we are at once aware of change and time. For even if it is dark and nothing is happening to our bodies but there is some change in the soul, at once time seems to have passed together with it. But also, when some time seems to have passed, some change seems to have accompanied it. So time is either change or related to change.] But, since it is not change, it must be related to change.

Having concluded from the previous discussion that time is neither change nor without change, and having embarked on a search for its
710,1 essence, he first establishes, in addition to the previous discussion, that we are simultaneously aware of change and time. For, he says, even if it is dark so that we neither undergo bodily change nor perceive other things changing, but there is some change in the soul, if the soul is thinking of something (which is the same, Alexander says, as 'have some awareness of change' and is clearly attention to change), 'at once time seems to have passed together with it'. Such is Alexander's interpretation, knowing that Aristotle does not mean that the soul is changing, but called the unchanging awareness of some or other change, change in the soul. But perhaps Aristotle thinks that psychic awareness is unchanging in a physical way, but envisages another type of change.

After the above he adds the correlative, that not only is one aware of time when aware of change but also aware of change when aware of time. So reasonably, if we take the two points together, time is either change or related to change. So if it has been proved that it is not change, it clearly remains that it is related to change.

[172] Literally, stades.

219a10-14 But since what changes changes from something into something [and every magnitude is continuous, change is consequent on magnitude. For it is because magnitude is continuous that change is continuous and, through change, time also. For] the quantity of time elapsed always seems to equal that of the change.

Having shown that time accompanies change, he next investigates how it is related to change. So he demonstrates that since change is continuous, it contains before and after. This is what time is, as we shall learn later. He demonstrates that change is continuous by demonstrating first that the magnitude undergoing the change is continuous: change is from something to something; what changes from something into something changes in some extended magnitude; so, since every magnitude is continuous, and change is in a magnitude, so change is continuous through the magnitude, as is time through the change. For the quantity of time elapsed seems equal to that of the change, just as the change seems equal to the magnitude. But if the quantity of change is continuous, like the magnitude (such as a mile's length), it is clear that the quantity of time is also continuous. For if the one were continuous and the other discrete, one could not say that the one is as great as the other. For such quantities are incommensurable. By neglecting to state one premiss, which says that what changes from something into something is in some magnitude, Aristotle seems to have introduced a slight obscurity into the text.

Eudemus also, in the third book of his *Physics*, seems to hold the same opinion, that change is continuous through the magnitude and time through change. He writes: 'Summing up we may say that time seems to be included among things that are continuous and have parts, and surely this holds also of change and length. Which of these then is fundamental, and to what do these attributes primarily belong? For it is not through their own particular nature that they have parts and are continuous. Change seems to be continuous and to have parts through being in a length, continuous through being in something continuous, having parts because, like length, it is divisible; for whatever part of a length be taken, it is also a part of a change. It is the same with time. We think that time, like change, has parts.' But Strato of Lampsacus says that change is continuous not only on account of length but also in itself, since, if it were segmented into stages, there would be an unsegmented change between the two stages.[173] 'Change', he says, 'is a certain quantity and divisible without end.'

[173] Strato probably argued similarly about time: although continuous, it does not *owe* its continuity to that of distance (Sorabji, *Time, Creation & the Continuum*, 378).

But if time includes departure from one present after another, whence it is called a change, and if it is not in this respect a quantity, that is not remarkable. For a number also divides, but dividing is not dependent on quantity; and a magnitude holds its parts together, but not *qua* quantity. For these are varieties of quantity, not quantity itself. So let departure from one present after another be a variety of change.

But what did Aristotle mean by 'for the quantity of time elapsed always seems to be equal to that of the change', since we call a swift change in a little time great? For no longer is the time equal to the change, since the former is little, the latter great. Or are the two sizes not seen merely in quantitative equality, but also in identity of relation? For the size of the relation of four to two is equal to that of eight to four; since it is double. In the same way, as the one change is related to the other, so time is related to time, since their speed and slowness are identical. Indeed Aristotle did not compare the two sizes through quantitative equality, but through the likeness of their continuity. For things not similarly continuous are incommensurable, as was said earlier. But let us not neglect to add to the subject-matter of the enquiry that the argument transferring the continuity of the magnitude subjected to change to change itself and time, and, from this, making the discovery of the essence of time, only applies to change of place. For other changes, such as alteration,[174] increase, diminution, coming to be and ceasing to be[175] may be changes of magnitudes, but not over an extended distance; so the argument catches not all time, but only that which accompanies change of place. But Aristotle apparently made his 712,1 approach from the clearer case, while Strato elegantly demonstrated that change as such was continuous; perhaps he also was aiming to ensure that the discussion should include not only change of place but also all the other kinds of change.

219a14-b9 'Before' and 'after' are primarily in place, [and there by position. But since there is a 'before' and 'after' in magnitude, there must be a 'before' and 'after' in change also, analogously to the other cases. But there is in time also a 'before' and 'after', since each of time and change follows on the other. The before and later in change is, as regards its subject, change, but its essence is different and not change. But we recognise time also when we distinguish the change, by

[174] *alloiôsis* is change of *quality*.

[175] It is surprising for Simplicius to count coming to be and ceasing to be as changes, even for the sake of argument. Aristotle held (*Metaph.* 1067b30ff.) that a thing's coming or ceasing to be was not a change in that thing. (J.O.U.)

> distinguishing the before and after. We then say that time has passed, when we perceive a before and after in change. We distinguish them by taking them to be different and to have another between them. For when we conceive of the extremes as different from what is between, and the soul says that there are two nows, one before and one after, then we say that this is time. For time seems to be what is bounded by nows, and we take this to be so. But when we perceive the now as one and neither before or after in change, nor as being itself related to some before and after, no time seems to have elapsed, since there has been no change. When there is a before and after, then we say there is time. For that is what time is, the number of change in respect of before and after. So time is not change, but exists in so far as change contains a number. Here is a reliable sign: we judge the greater and the less numerically, but change by greater or lesser time. So time is a kind of number. But since number has two senses (for we call both what is numbered and numerable and also what we use to number it a number), time is what is numbered and not that by which we number.] What we number and what we number by are different.[176]

The whole aim of what is said is to show that just as change derives its continuity from the magnitude of that in which the change occurs, so the change likewise takes on from it the before and after that belongs to a magnitude through position, and similarly time from change, as does also continuity. For in magnitude the before is, for instance, the part nearest to the competitors, the later the end of the track, in motion the before is the motion at the first point, the later that at the later point. But as regards magnitude, the before is always co-present with the after (for the first and last points co-exist), but as regards change the before is annihilated and the after takes its place.

It is, however, not the same thing to be a change as to be before and after in change, just as it is not the same thing to be a magnitude as to be before and after in magnitude. But, even if they are in their substrate the same, they are different in account. Just as five things may in substrate be for example pieces of wood, but their being five is a numerical fact. Also the substrate of change is

176 Aristotle is traditionally translated as holding that time is 'the number of motion'. But Aristotle is concerned with all kinds of change, whether motion or other, so 'number of change' is preferable. Even this translation is defective since 'number of' with a singular noun is in English illiterate, except in such irrelevant uses as 'the number of your house'. What Aristotle wants to maintain is that there is time because in change we can recognise nows or instants of it as being first, next and next ... Simplicius will offer his explanation in what follows. (J.O.U.)

the actuality of the changeable, but the before and after in change is a number and ordering in such a thing. Similarly in the case of a magnitude in which there is a before and an after in change, the substrate may perhaps be columns, but the before and after is their arrangement. This is what he means by 'as regards its substratum[177] it is change', i.e. its existence and substrate. But its essence, i.e. its account, is something else and not change.

Having thus separated the before and after in change from change, it remains for him to investigate the essence of being before and after, through which they differ from change. It is like the case of a line: so long as we think of it as one and continuous, we do not think of the portion of it that is before in position, and that which is after, but when we divide it and distinguish points, we at once make one before in position and one after and treat the extremes as other than the middle of the line that they include. It is the same with change: we do not think of the before and after in it except when we divide and distinguish it into, perhaps, two stages that we perceive as different things. How, then, do we distinguish? Presumably not by a point, since change has no position, but in some other way. For I call 'before' the change that I have undergone when, perhaps, walking, 'after' the change that I shall undergo. So the distinction is made by one having passed and the other being future. But the past and the future are time. So time is the same thing as the before and after in change, when they are distinguished. They are distinguished when the nows strike the soul as two, one as the beginning of change, the other as the end. For then we at once conceive of the time between the nows as different from them, and often we may take one now as the end of what has passed and as the beginning of the future and in that way again we gain a conception of time as both before and after. I think this is a better way to understand 'as being itself related to same before and after' than as Alexander explained it, saying: 'the now is one and the same, but is in some before and after when we think of it as being, as it were, moved along and occurring sometimes at one, sometimes at another, stage of the change.'

So, on all counts, the concept of time is gained by taking separately the before and after in change. They are taken separately when they are taken as different entities, i.e. when numbered. For number is not found in what is one and altogether the same, but in what contains different things. So time is the number of change, since it can be numbered in respect of before and after. It is also clear that it is the number of change, since every greater and less is

[177] The expression here is not *hupokeimenon* as in the surrounding lines (translated as 'substrate'), but *ho pote on*, an abbreviation of *ho pote on khronos esti*, roughly: what (else) it has to be in order to be time.

judged numerically. But the greater and less of change is judged by time. For there is more change in more time, if taken to be at constant speed, and less in less. So the conclusion is drawn in the third syllogistic figure that time is some numbered amount. Alexander also draws the conclusion in the third figure from two affirmative premisses as follows: the greater and less of everything is judged numerically, so clearly of change also; but the greater and less of every change is judged by time; so time is some numbered amount. But we must pay attention to whether we have the same subject-matter when we say 'every more and less is judged numerically' and 'the greater and less of every change is judged by time'. It is clear that 'in respect of before and after' is necessarily required, since many changes can be counted simultaneously, as, in your case, that you are growing, and growing warm, and moving on from place to place, and thus undergoing three changes. But time is not that sort of number of change, but that in respect of before and after.

It is in no way surprising if, holding time to be what is numbered in change, he did not say 'that which is numbered' but 'the number'. For that which is numbered is called a number, just as what is measured is called a measure; for example, abstract[178] number, the number of cattle, the number of horses. Again, there is a wooden quart and one of corn, a gallon jar and a gallon of wine. And we eat the quart and drink the gallon, but not the wooden one or the clay one, but of corn and wine. Also the number of numbers would not suit time, for that is discrete and not continuous. But what is numbered can also be continuous, like the eleven-foot spear. Also, 'what is numbered' is ambiguous, in one sense being in respect of quantity, as when we speak of two or three changes, in the other in respect of order, as when we speak of the former and the later change. So he wants to demonstrate that time is the number of change, both as numbered and as numbered in respect of order. This is its peculiar property, for number as such is common to change, cattle and horses. So, when he said 'for time seems to be what is bounded by a now, and we take this to be so', he was showing how time is the number of change, that it was so as the numbered and not as what numbers. For what is numbered in change, and so determined that the before and after were different things, was to be in change but not change itself, since the before and after have been shown to be different from change. For time is this number of the before and after in change, not of change itself. For it is the number of changes, as has been said, and if many people are changing at once, e.g. my walking and getting hot and growing, or the parts of

[178] *monadikos*: made up of ideal units.

the change themselves, yet this number is not time; but the reckoning up of each change in these many cases by reference to before and after – that is already time. Nor, also, is the before and after in magnitude time, since both persist in it; but those in time do not persist.

But, since we seem to measure by time (for we say something lasts so many years and the festival so many days), on this account some people alter the text, as Aspasius says, as follows: 'time is not what is numbered but that by which we number.' But it is plain that time also is measured by abstract number; for we say how many times have passed. But perhaps there is nothing to prevent time being numbered and also numbering. Aspasius also sets out the next part of the text as follows: 'there is a difference not in that by which we number but in what we number.' He also says that 'if we accept that 715,1 properly we number by abstract number, we may say that that by which we number cannot admit difference; for we repeatedly take the unit, but a unit does not differ from a unit in genus, in species or numerically. For what differs in number will differ either in form, as Plato does from Socrates, or in order, as point differs from point and now from now by being before or after, or position, as above and below, or as taken in some other similar way, which is not the case with units. It is better', he says, 'to take the now as that by which we number.' But perhaps that which numbers is the same everywhere in manifolds. That is the aim, then, of what he says.

While Alexander explained almost everything well, he added a few small points also that deserve attention. For Aristotle, at the beginning of the passage now before us, said that 'before and after are primarily in place'; Alexander says that this is equivalent to 'before and after are primarily in place, i.e. in the moving magnitude'. 'For', he says, 'it is by position that the magnitude that is in a place has a before and after.' What did he mean by this? For the before and after by position belongs to place, but not to the magnitude that is in a place. For the positions of a magnitude are relative to motion, which has its positional before and after from place. Also its positions are not simultaneous, but the magnitude occupies different ones at different times. But the positions in place are simultaneous. How, then, could the before and after in motion be derived from the before and after in position of the thing that moves? So perhaps Aristotle, when he said that there is a before and after in magnitude, there too meant not 'in the moving magnitude' but 'in the distance over which the motion occurs'.

With regard to Aristotle's statements, one may be puzzled (1) why he places time in the before and after of change of place, as is clear from the transition from the before and after in place, which has position, to the before and after of motion – and yet time is seen not only in change of place but in all change; (2) how it is that, having said 'since there is a before and after in magnitude, there must be a before and after in change also, analogously to the other cases', he said that time was that before and after occurring in change through likeness to magnitude. But this before and after would be in position, and not in time, since the one change requires a magnitude that is before in position, the other one that is after. Thus the first stage in motion from Athens to Thebes is in a position near Athens. [716,1] But if there is also a temporal before and after in the change, perhaps that is not derived from magnitude. In fact there seem to be two types of before and after in change, one derived from place and one from time. And, as he said previously that the idea of place came from the before and after in position of that which is in motion, perhaps he means that we get the idea of time from the before and after in change, not as from the position of the magnitude entering into change, but as from its extension in time. For, since change is central between place and time, he adds a measure or number of before and after from both. A thing gets its place from place and its time from time. It gets its place in so far as its extension has a position, and it gets its time in so far as its existence is in flux. If it had position only it would have had no need of time, for it would have its being at the instant, and if it was pure flux it would have had no need of place. For place was the measure of its extension as regards position.

So if someone were to define time as the number of the before and after with regard to being the flux of change and similarly place as the number of the before and after with regard to the position of the extension, I do not think he would be altogether wrong. But since there are two sorts of number, the one enumerating or enumerated in quantity, when we say 'one, two, three', the other ordinal, as when we say 'first, second, third', time is this latter sort of number; that is why it is distinguished into before and after. How then is the definition defective, if before and after in respect of flux or the continuation of existence is not added to distinguish it from before and after in position? Perhaps before and after in position is only contingently in change; for change seems to be extended and to have a position because of the position in the interval of what is moving in the interval; that is why it passes on. But its extension in flux is the essence of change. Aristotle also, making the distinction later, says that the existence of everything is measured by time.

So perhaps it would be clearer not to characterise time from the before and after of place, nor to say in what follows that change is equal in quantity to magnitude, and time to change. For these give the extent of change in relation to position, not in relation to time. But we should attend specially to his statements where he says that the extension of existence is measured by time. So much in reply to the second puzzle.

In reply to the first we say that even if all change is measured by time, still change of place is the most obvious of all, so that the time that accompanies this change is more obvious than time accompanying any other kind of change. Perhaps, if one examines the matter more closely, all the other forms of change also, such as growth, coming to be and alteration, are measured by this time that accompanies change of place; for we measure the duration of these other changes by hours and days and months and years.

Eudemus, however, in the third book of his *Physics* wrote as follows: 'One might raise a problem about time, of what or of what sort of change it is the number. For it does not seem to be the number of some one change, but equally of all. For it is equally the number of everything that has a before and after; and that is common to all change. Just as the knowledge of a mortal animal is not that of a horse, or of a man, or of any other species, but universal, so time also is not the number of some particular change, but equally of all (for every change whatsoever is[179] everywhere and equally); but all the same it is primarily and specially the number of the prior and more uniform change. For it becomes more familiar in definite cases.' Seeing that Eudemus has written that so clearly, it is worthwhile to consider what this universal is. If universal change actually exists, why not man and horse and each of the other species? But why is there not a time in general and a species of time, just as there is change in general and a species of change? But if the universal of change is either nothing or *post rem*,[180] it will be the same for time as for change, and this [universal] time will be either nothing or merely a concept.

But since it seemed to Alexander that the definition of time given by Aristotle had not the same content as what Plato said about it, let us then compare these two accounts. For Aristotle said that time was the number of change in respect of before and after, but Plato says that time is 'an eternal image proceeding in accordance with number'[181] of the eternity that remains static in unity, contrasting 'in unity' with 'in accordance with number', the changing with the static and the image with the paradigm. Now observe that Aristotle best of all men well understood Plato's conception of time. For what is this 'eternal image proceeding in accordance with number' other

[179] Reading *esti* for *estê* with Spengel and Diels; but the clause in brackets seems false and ungrammatical as so emended. If we accept Diels's hesitant suggestion that *an* is the remnant of *arithmêtê* we get the better 'for every change everywhere is numerable'. Possibly we might read *esti dê pantakhê kai homoiôs arithmos hêstinos kinêseôs* – it [time] is indeed the number of every change equally everywhere. (J.O.U.)

[180] Dependent on instances.

[181] Plato *Timaeus* 37D. (J.O.U.)

than time existing in accordance with number that is neither paradigmatic[182] number nor abstract [cardinal] number, but that which is viewed as an image concerned with the order of change, i.e. the before and after?[183] For the number belonging to the changing *qua* changing has at different times a different before and after and both introduces it and determines it in regard to order.

Aristotle saw this well and, giving a clearer explanation, called time the number of change in respect of before and after. So, if 'in accordance with number' expresses 'in regard to order', as Alexander suspected, it expresses nothing different from 'in respect of before and after'. For this is the ordinal number, just as the abstract is one, two, three. But see how well Aristotle grasped the eternity (*aiôn*) as conceived by Plato. For while Plato says that the animal itself[184] preserves that which is everlasting (*aïdion*) on account of eternity, Aristotle says that it[185] (*aiôn*) has got its name through being for ever (*aei ôn*), well interpreting that it preserves and encompasses the whole of time by being static in unity and that from it being and life irradiate all things. But what first partakes of life is the first living thing.

The great polymath Galen objected to some of the things said at this point, saying that time revealed itself, so let us consider this objection also: 'Having enumerated many things that are signified by "before" and "after", he says that the rest do not fit with the definition, but only the before and after in time. So time is the number of change in time.' Themistius proposes to eliminate this objection by two arguments, saying 'One must know that the before and after in change are not respectively before and after on account of time but rather themselves create the before and after in time. They come about from the before and after in magnitude and in position, through which they also possess continuity. That is why Aristotle explicitly says, "The before and after is first in place, then in position, then in magnitude, and necessarily also in change".' In reply to that Galen would say that the before and after in change following on that in the magnitude over which the change takes place is chiefly in position; for the before and after in magnitude was

[182] *paradigmatikos*: the non-composite Platonic Form of Number, as opposed to number made up of ideal units (*monadikos*, or abstract number).

[183] Diels has a full stop here instead of a question mark. (J.O.U.)

[184] The animal itself is the Platonic Form of the living universe, called in lines 11-12 the first living thing. It is itself eternal or timeless (*aiônion*), and by means of eternity preserves its copy, the living universe, which is merely everlasting, or existent at *all* times (*aïdion*).

[185] There is a change of subject here from Plato's eternal Form to eternity itself, but Aristotle's point about eternity (see *de Caelo* 279a27) is treated as a parallel one: as Plato's eternal Form preserves the everlasting universe, so, according to Aristotle, eternity itself preserves time.

of that kind. As has been said, the before in time is different, accompanying the before and after in change, but not having its
719,1 before and after through position, but through the extent of its existence; and this is something other than the before and after in position derived by motion from magnitude.

Introducing the second reply he says: 'But let it be conceded that before and after signify nothing beyond the temporal, as he thinks. What absurdity arises from that? For we ourselves say that time is nothing other than the before and after in change. But there is every need for definitions to signify the same as the names. So he fails to notice that he is censuring that in the account on account of which he should most accept it.' In reply to this Galen would say that if there were but one before and after in change by temporal succession it would be well said that time was this before and after in change, and time would not be something self-revealing. But if there is one, that in position, which is not temporal, another that is temporal, one must say that time is that before and after in change that is in time. But perhaps if one were to take the before and after to be that in the extent of existence, and not that in position, time is signified and not self-signifying. Also its name will not be simply the same thing as the before and after, but as that sort of before and after which is time.

Having said that there are two meanings of 'number', he lists three, the numbered and the numerable and that by which we number, so that it is clear to all that the numbered and the numerable are the same.

> **219b9-10** And just as change is ever different, so is time. All simultaneous times are the same.

Having shown that time is related to change, for the numbered in change is the before and after, he confirms their kinship also from the flux that is common to both change and time. For just as change is perpetual succession, the first not remaining but ceasing to be, so too the before in time is continually ceasing to be as the after succeeds. It is clear that it is not only the temporal element in change that does not remain, but also that of position, for here, too, the before ceases to be and the after succeeds. However, in the magnitude itself in which the change occurs, the before and the after remain simultaneously. For it is only in things that have their being in becoming that the after succeeds as the before ceases to be.[186] So
720,1 the before and after in change, not that according to its position but that according to the extension of its being, is time, not *qua* change,

[186] See Aristotle *Physics* 3.6, 206a22-4.

but *qua* being a before and after of that sort. But there is a before and after in change according to position, by which as one walks from Athens to Thebes the first change is that in the Athens area. This before and after, even if it has its being in becoming, still, regarded in relation to the position, which remains still, seems itself to remain still in a way, when the change itself is regarded as extended together with the magnitude over which it occurs. But when it is seen as change proper in relation to flux and having being in becoming, then it becomes temporal, having a past and a future.

Having stated what is common to time and change, he reasonably adds their difference. Not only is the earlier change other than that which succeeds it, which is a feature shared with time, but also simultaneous changes differ from each other, some in kind also, as if one thing were to change place, another grew and another altered, some only numerically, as if several things changed place at once, or grew or altered. For the change is in that which changes, so that there are as many changes as things changing; but time is everywhere the same for each thing referred to the same instant, and one time does not differ from another. Also a difference between those things present at a time, whether in kind or numerical, is not a difference in time, for events in Athens and in Corinth are at the same time. This is because, while each changing thing has its own proper change, time is not thus the number of each distinct change, but of all as one. For the before and after as such is the same in all changes that occur simultaneously. For while one change is in this place, another in that, and while one is a local motion, another a growth, another an alteration, it is not as such that they have a before and after, but a common one as change. That is why they are called contemporary, because they are simultaneously at the same time, just as the number five is the same, even if applied to different groups. The only chronological difference is that through the difference of the nows. For time has its being in the taking of these nows, when one is taken as before, another as after. But if the difference in the now creates the difference in time, then also the difference in time will be only in being before and after, while there is no before and after in simultaneous time. So there is no difference in simultaneous time.

I think that someone might raise a difficulty about this, saying that if time were number as that which numbers it might perhaps be one and the same. But if it is so as the numbered, and this is
already instantiated and applied, how can it be the same?[187] But if, 721,1
taken generally, *qua* time it is the same, why should not change be

[187] Diels has a full stop, not a question mark, as he has at the end of the next sentence. (J.O.U.)

one and the same *qua* change? But if change also has both a common application and one to individuals, then, first, those who say that the common is either nothing or *post rem*[188] will ask what this common application is; secondly, why has not time also a common and an individual application?

I think that one might solve the problem by saying that time is not the number of every change according to the present theory, as of both my walking and your becoming white. If it were, it would not be undifferentiated and simultaneous as distinguished at a single now. But time is the numbered in the first and simple motion of the heavens, and numbers the befores and afters in different changes according to one now of its own, that is, according to the before and after of that motion. That is why there is the same time everywhere. Also Aristotle himself, as he goes on, will make the now not only numbered but also that which numbers when he says: 'As a limit the now is not time, but characterises time; but it is time as numbering.'[189] If this is so, it will be undifferentiated as that which numbers, but different as numbered. But perhaps, more in accord with Aristotle, we will say of the different changes, that, even if they differ from each other either in species or numerically, still the common feature of change, to have its being in becoming and to be ordered, in accordance with which their before and after that is found in the continuation of their being is determined, as is time also, is one and the same everywhere. For the before in my walking and of your becoming white, as change, if they are simultaneous, will be in the same now. Similarly the after will be in the same now, as is made clear towards the end of this book of the *Physics*. For even if the common feature of change is not transcendent and independent, but instantiated, though still retaining its generality, it will be everywhere the same. For in that respect changes will not be separated, even numerically.

219b10-33 The now is identical as substrate (*ho pote ên*), [but its essence varies. The now measures time *qua* before and after. The now is in one way the same, in another way it is not. As ever differently positioned, it is different, for that is what it is to be a now, but as substrate it is the same. For, as has been said, change accompanies magnitude, and time change, as we

188 For *post rem*, see n. 39 above.

189 The commonly received text of Aristotle's *Physics* reads at 220a21: *hêi men oun peras to nun, ou khronos, alla sumbebêken: hêi d'arithmei, arithmos*, which presumably means 'as a limit the now is not time, but characterises time: but as numbering it is a number'. Simplicius omits the word *arithmos*, which gives a different sense to the passage. Ross, in his edition of the *Physics*, marks the passage as corrupt. Philoponus has the same text as Simplicius. (J.O.U.)

maintain. Likewise, the moving body accompanies the point, and by it we recognise the motion and the before and the after that the motion contains. But this substrate is identical, whether it be a point or a stone or something else of that sort, but the account of it is different, in the way that the sophists consider that Coriscus in the lyceum is different from Coriscus in the market-place. Thus the object is different through being in different places; but the now accompanies the moving body, as the time accompanies the motion. For it is by the moving body that we recognise the before and after in motion, and since the before and after is numerable, there is a now. So in these cases also the now as substrate is the same, for it is the before and after in change, but its essence is different, since the now exists in so far as the before and after is numerable. This is especially obvious, for there is change because of that which changes, and the motion because of what moves. For what moves is a particular thing, but change is not. So in one way the now is for ever the same, in another way it is not the same,] just like the moving object.

Having said that time is associated with change in coming to be and ceasing to be, and again that it is different, because changes are different from each other while 'all simultaneous times are the same', he says that the explanation of this is that the now is identical with the substrate (for that is what *ho pote ên* means). For, as he said of the before and after that 'the substrate is a change, but its essence is other and not change', so he has also said of the now that in substrate it is one and the same but differs in account, in so far as one is before, the other after. In this way the before and after have their being, and the now is a measure of time by which time is 722,1
measured as successive, just as number is measured by the unit. Reasonably, then, the difference and identity of time is derived from the difference and identity of the nows. Time will be identical as determined by the nows taken as identical, even if they are within different changes, just as today is the same when one thing is growing, another grows white, another takes a walk, since it contains the same nows at the beginning and at the end. But if the nows are taken as different, since some are before, some after, the time will be different. So simultaneous time, determined by the same undifferentiated nows, will not be different.

The text seems to contain a slight unclarity, because Aristotle, contrary to his custom, explains the same thing in different ways, or, rather, clarifies the earlier by the later. For having said that the now as substrate was the same, but its essence was different, he adds that the now measures time, *qua* before and after; then, wishing to clarify how the now is different, he again takes the distinguishable whole, saying that 'the now is in a way identical, in a way it is not; for, as at one time and another, it is other', which is to say that it is both the first and the later. But then he again adds that 'the substrate is the now, the identical' which is the same thing as 'the now is identical as substrate'.

But, I think, since it is not indisputable that the now is identical with the substrate but different in account through it being before and after when taken successively, he tries to strengthen it by its connection with what goes with it. For previously change was said to accompany magnitude and time to accompany change, because the quantity of change was equal to that of the magnitude in which it occurred, and the quantity of time that was the numerable in change was equal to the quantity of the change. Since these things accompany each other, the things that originate them will accompany them also. But the point originates the magnitude, taking a line to be the flow of a point, and a line is the first magnitude, and the moving thing originates the motion, and the now originates time. As, then, the same point, by flowing, creates the magnitude, but is not set alongside another point, and as the identical moving object creates the motion, so the now, that is the same *qua* substrate, being numbered as before and after, defines and creates time.

He confirms that this is the case with the now from the clearer case of the changing thing, and he reminded us that this is clearer
723,1 because through it we come to know change and the before and after within it. For it is because the moving object comes to first one and then another part of the substrate that we recognise both the motion and the before and after, both in change and in time. He takes it as obvious that what is changing is identical in substrate, whether it be a line or a stone or anything similar. That the stone remains the same thing is very plain, even if it is in different places at different times. That the point remains the same in its flow is clear, since the event is continuous and is not composed by adjoining points. But the next appears together with the former that remains, through the flow of the single point. But in account the moving stone becomes different through coming to different places within the magnitude within which it is moving. That this seems to be a sort of difference he confirms also from the sophistic arguments which, by transferring this difference to the substrate, seemed to draw an absurd conclusion that the identical thing became different from itself; for they said that 'the same Coriscus is sometimes in the market-place and sometimes in the Lyceum. He who is sometimes in the market-place, sometimes in the Lyceum, becomes different from himself.' They did not take notice of the way in which 'different' was

taken, but rather transformed contingent into essential difference. So he showed that it was sophistical to think that through the before and after being different in account they were therefore different in substrate.

Having shown how that which moves is identical with the substrate, though different in account, he says that the case is the same with the now, since time follows on change and has the same features, as was said earlier, and similarly with the now and what changes. The likeness and identity of the cases is shown by the relation of the now to that which moves, and we recognise the before and after in change in no other way. For these would not have been distinguished if the moving object had been at rest, but, since the moving object is once at the previous and once at the later place,[190] through that the before and after of change comes about. In so far as this before and after is numerable, in that way, he says, the now is displayed. It is clear that according to my opinion this before and after in change is to be taken not as being in position but according to the extent of being, so that the now also is the same in substrate, but different in account. For its being numerable and distinguished as ever different gives to it the difference in account.

Eudemus also says the same about the now in the third book of his *Physics*, writing as follows: 'If we think of a moving point which itself becomes continually different through its being in one and 724,1 then another part of a line, we may think that the same holds with the now. The substrate, however we are to think of it, remains the same, but by being at different times it will itself be continually different. So that in substrate, whatever it may be, it is the same, but different by being at different times. In so far as the now is before and after, in that way it is different, and in that way time is a number; so that when the nows are the same, so the number and the time will be the same. When the nows are different, the rest will be also.'

Having shown from the before and after, the dependence of the now on the changing thing, he shows it also from the fact that both are recognised in the same way, change through the changing thing and time through the now. For change is recognised by the changing thing and time by the now, which seems to exist only through the presence of time. He made clear how the moving thing is more familiar than motion, saying that 'what moves is a particular thing, but motion is not'. For the whole of the moving object is simultaneously a substance and an existent, but change is a characteristic and has its being in becoming; for change exists in the change of that which changes.

190 Reading *topon* for *tropon*, which I take to be a misprint. (J.O.U.)

Having said 'for change [is known] through that which changes', he added 'and motion through that which moves', not because motion is not a change, but because in the case of motion it is more
20 clear that motion is known through the moving object. For even though alteration also is known through that which is altered, and growth through that which grows, since each of these is a substance, none the less motion is known more clearly from the moving object, since motion is more obvious than the other changes. Finally, drawing a conclusion from what has been said, he says that 'in a way
25 the now is identical, in a way it is not, and the same is true of the moving object'. For, from its dependence on the moving object, and through having its being from it, the now also has its identity and diversity. At this stage it seems to me to be worthy of consideration in what way he says that the before and after is identical in substrate. For how can what is past be identical with the present
30 and that which is not yet with that which is? For the moving object, though coming into different places, can correctly be said to be identical in substrate, since it remains in existence, but how can change and time and in general everything that has its being in becoming be identical in substrate? Nothing prevents one from seeing that in the case of the point; for by its flow the magnitude is
35 created and remains as a whole, and does not have its being in
725,1 becoming as does time. But how did he show that the now accompanies the moving object and the point in and through these cases, since there is such a difference between them, since the one remains, the other does not. How, then, can what does not remain have identity of substrate, like what does? Secondly, I want to
5 enquire, if there are these three, magnitude, change and time, mutually dependent, as he says, magnitude being limited by points, time by the now, what is analogous to these, I mean the point and the now, in the case of change. For there should be some other limit to change and not the changing thing. For what relationship has this to the now and the point?

10 Perhaps one should reply to the first of the problems that the nows differ from each other in no way save that one is before, the other after, and since in neither genus nor species, so also not in number, as unit does not differ from unit. For numerically different things, as has been said, have some different characteristics, as has Plato from Socrates. So perhaps they are said to be identical in
15 substrate because they are in no way different. But the comparison with the point and the stone seems to be forced, since these continue, the others have their being in becoming. So, if we do not niggle about this comparison, I think that it is better to say that 'identity in substrate' means the same as 'identity in species'. For Aristotle himself will say as he goes on that just as change is the

same in species through repetition, so time also. But to be the same in species in such a way as to admit no difference and to be numerically identical seem the same in a way, as was said earlier in the case of units. A little later I shall add Eudemus as a witness. For these reasons the comparison of the now with the point and the moving object does not seem altogether harsh. In reply to the second enquiry one must say that a movement is the limit of motion, but is not the principle nor the explanation of change. So when he postulates the point and the now as causes, the one of the line, the other of time, he reasonably took the moving object to be similar to them.

> **219b33-220a4** It is clear both that if there were no time [there would be no now, and that if there were no now there would be no time. For as the moving object and the motion are conjoined, so are the number of the moving object and that of the motion. For time is the number of motion, but the now is like the moving object,] as a unit in a number.

Having said how the now is related to time, that it is so related as is the moving object to motion, and that the now is more familiar than time, he now displays their coexistence, once again by analogy with the coexistence of the moving object and motion. For, he says, just as the moving object and the motion are conjoined (for it is impossible for the one to be without the other, if you take the moving object *qua* moving and not as a substance), so are the now and time. Also he demonstrated the necessity of the connection from the analogy of the connection of the moving object and motion. For the now is the number of the moving object, like a unit in harmonising indivisible to indivisible. But time is the number of change. For just as the unit, by being repeatedly taken, creates the number, so the now creates time and the moving object creates motion. But time is the number of motion, not *qua* motion, as has been said, but through the before and after within it. For time is not the number of change itself, but is the interval between the before and the after which by repetition create the interval. But the unit differs from the moving object and the now, in so far as it makes number discrete, but they make both change and time continuous.

> **220a4-21** Time is both continuous through the now [and divided by it. This too corresponds with motion and the moving object. For both the change, i.e. the motion, is one on account of the moving object, since that is one – not as substrate, which might pause, but as moving – and this distinguishes the motion as before and after. This corresponds in a way to the case of the point; for the point both makes the length

continuous and divides it. It is the beginning of one bit and the end of the other. But when one takes it thus, using the one point as two, one must pause, if the same point is to be beginning and end. But the now, through the motion of the moving object, is always different. So time is a number, not as of the same point, which is both beginning and end, but rather as the end points of the line are a number. Also it is not as the parts of the line, both because of what has been said – for one may use the middle point as two, so that there will be no change, and also clearly because the now is no part of time, nor the division any part of a change, nor a point a part of a line.] For the two lines are parts of the single line.

He again demonstrates the continuity of time and its division at the now from its connection with the moving object and motion. For motion (*kinêsis*) also, i.e. locomotion (*phora*), becomes continuous and one through the moving object, and in no other way. It becomes one and continuous through this because this also is one. This unity is double, both in relation to the substrate, which once again he called *ho pote on*, as if, for example, the moving object was a stone, and through its being the actual moving thing, i.e. through its account as the moving object. But it will not be the cause of continuity through being the substrate. For it can pause in its motion, and what is a stone will continue to exist, but that will not make the motion continuous, since, though it continues as such, its motion is cut through, being brought to a stop. But if it is taken as one through the account of a moving thing and as being a moving thing, then it becomes the cause of the change being one and continuous.

But the now does not only make time continuous, in accordance with its connection with the moving object, but it also divides time, distinguishing the before and after of change. For that is how time is divided. It resembles not only the moving object in producing continuity and division, but also the point, in a certain fashion. For the point, also, taken as identical, is 'the beginning of one bit and the end of another', and it makes the length continuous, since it may be taken as one and the same in each portion of the length, but none the less divides it. For the division of a length is at a point. In a similar way the now both makes continuous and divides time. As regards continuity and division the comparison with the point is complete, but, in so far as the identical point is beginning and end, we treat the one as two. Thus there must be a stop at the one point taken as two, at one time as limit and at another as beginning. For it is essentially at rest and in a position. But we do not take the moving object twice; for it is always in one state after another. For if it were twice at the same place it would have to be at rest; but it is presupposed that it moves continuously. Similarly it will not be possible to take the now twice. How then is there time in virtue of the numbering of the now? It is surely not as the same point functions twice as beginning and end, but rather as this and that are the two limits of a line. These, being two different things, determine the interval of the line between them. In the same way the now is numbered as one and another, one before and the other after. There is no way that this can be taken twice, for time would be stationary, which cannot be. But there is nothing absurd in taking the point twice, since it has a position and is at rest. But the now does not rest so that it can be taken twice. It is clear that if a point were taken as limit and beginning the line would be divided at it. But when divided it is no longer continuous; so nor will the change in it be continuous. But time is continuous since the change is also continuous. So time is the number of the now, but not through the same now being taken twice, like the point that is taken twice, but as it is first one and then another, one as beginning, the other as limit. Also they will not be beginning and limit as portions are often said to be first and last.

He proved that the nows are not like the point taken twice by saying 'for one may use the middle point as two, so that there will be no change'. Presumably if time does not cease to change, the now within it is not like the point that is taken twice, but as beginning and end. He next confirms that it is not beginning and end in the way that parts are, but as limits, once again through its likeness to time's ingredients. So 'because of what has been said' etc. is the explanation of 'not as of the same point, which is both beginning and end, but rather as the end points of a line are a number'. Similarly 'and also clearly because the now is not a part of time' etc. explains 'not as the parts of the line'.

If, indeed, the text read as Alexander and Aspasius write: 'and it is also clear that time is no part of change, as the point is no part of 728,1
the line', in that case they say that by time he means the now, which is a part neither of change nor of time itself; for it is a limit of time and a limit is not a part. For also a point is not a part of a line, since a line is a part of a line. That is Alexander. But Aspasius says: 'Time is not a part of change because it numbers it by the now, and that is a limit of change but not a part.' But how could the now be said to be a limit of change? For where the substance of the thing is different it has different limits. So if time is different from change its limits will be different also.

So I think that the reading of the majority of the manuscripts is better, which was clearly known to Porphyry and Themistius and which runs: 'and also clearly the now is no part of time, nor the division any part of a change, just as points are not parts of a line. For the two lines are the parts of the single line.' For it is clear that lines are parts of a line, not points. He calls the limit of a change a division, as that is how it is divided, which is what he calls a movement in the discussion of change.[191] Where Aristotle says 'So time is a number, not as of the same point', Alexander says that he has called the now time: 'So,' he says, 'the now is not a number as the number of a point, as when we take the one as two and twice.' But perhaps it is not necessary to take time to be the now, but, as he says, time is a number composed of unitary nows, the same one not being taken twice, but one after another.

> **220a21-6** So the now, *qua* limit, is not time, but a characteristic. [*Qua* numbering, it is a number. For limits are of that only of which they are limits, but a number is of these horses, say ten, and belongs also elsewhere. So it is clear that time is the number of change in relation to the before and after] and is continuous, being of the continuous.

He has said that the now is a limit in the same way as points that are the extremes of a single line. Since the now is as such the limit of time, and since time is nothing other than what is numbered in change, in relation to the before and after, which are the nows, he takes notice that, when the now is taken as what is numbered in
729,1 change, then it becomes the limit of change and is thus contingently a limit. Also then the now is treated as the substrate, and is in that alone of which it is the limit. But, when it is treated as the limit of time, this is not so contingently; for as such it belongs to time. And since the now is a number everywhere, and a number everywhere is a number that numbers (for the ten that counts ten horses also counts ten men), so the now is simultaneously everywhere and not in the change alone, of which it is a limit, not as a now but as a movement.

Alexander enquired what he meant by the now numbering (for time was not a number as numbering but as numbered). 'Either', he said, 'the correct reading is not "numbers" but "numbered", or the nows are numbered in change but number time. For the divisions of time are at nows, and we number time by these, so that the now is numbered as being in change, but numbers in relation to time.' That Aristotle takes the now as a number that numbers he makes plain by saying that 'a number is of these horses, say ten, and also belongs elsewhere'.

[191] The word *kinêma*, here translated 'movement', occurs frequently in Book 6 of the *Physics*. The translation is *faute de mieux*. (J.O.U.)

It is possible to see change also as numbered by the now. For if time is the number of change in relation to the before and after, and the before and after are nows, change is numbered by the now. But since time is not the nows, but what is between the nows, which do the numbering, it itself becomes numbered. It is clear that every participated[192] number could be said to number in this way, like the half score of horses; for horses are numbered in that way. But such a number would not be said to be everywhere, for it is that which is numbered. So if the now were said to number, that now would be said to be the numbered in primary change, but what numbers other changes: for that is indeed universal.

With this the discussion is complete. He adds that time is continuous, because it is the number of the continuous. For this is the essence of time, to be the number of change. For the being of time extends together with the continuity of the numbered change, and so must itself be continuous. Since change is not interrupted, it could not be interrupted. For since time divides by the limits of change, not *qua* limits but *qua* before and after, which makes it time, it must itself be continuous.

CHAPTER 12

> **220a27-32** The least number, [without qualification, is two.[193] But a particular number may or may not have a minimum. Thus the least number, in quantity, of lines is two, or one, but there is not a least in magnitude. For every line is divisible endlessly. For the least number in counting is one, or two,] but there is not a least number in magnitude.

Having said that time is a number and continuous, he wishes to resolve the apparent conflict in this statement. For a number is a discrete quantity, but the continuous is not discrete: and, as he himself says, 'in number there is a least, e.g. two'.[194] For two is the first number, and is not divisible into numbers. But in a continuum 730,1 there is nothing that is least; for the continuous is divided into continuous parts without end. So, if there is a least number, but there is not a least time, since it is continuous, time should not be a number. Resolving this problem, he makes use of the earlier statement that time is a number, not as one that numbers, but as one that is numbered. Two things are called number, one without qualification and primarily; this is that with which we number, and

[192] Instantiated.
[193] One is contrasted by the Greeks with number.
[194] This is not an exact quotation from the received text, as above. (J.O.U.)

it is discrete and composed of indivisible units. Therefore there is a least in it. But number as that which is numbered, is not referred to
10 without qualification, nor in its primary sense, and exists in continua. But, taken as a plurality, this also has a least, as two lines or two years are the least; but this is not so in magnitude. For here we have quantity in magnitude and not only in plurality. Alexander well comments that 'he said that two was the first number, and not one,
15 because one is not yet a number, if a number is a plurality consisting of units, and the unit is indivisible. But he took it that one line and one time could be said to be the least in number. But this is either said generally, because the argument is not now concerned with such matters, or said because in the case of number as that which numbers the unit is entirely indivisible and not yet a number as plurality, but
20 in the case of the numbered one is already a number; for there is a compound in it of unit and substrate.'

> **220a32-b5** It is also clear that time is not said to be quick and slow, [but is much and little, long and short; as continuous it is long and short, as a number much and little. But it is not quick and slow:] for no number with which we number is quick and slow.[195]

25 Having shown how time as both a number *qua* numbered and continuous *qua* magnitude has a minimum as number, but not as continuous, he draws the consequence that time reasonably has a much and a little as number, but a long and a short as magnitude. These are in each case properties. But it is not also said to be quick
30 and slow, because no number is quick and slow. For neither the abstract number with which we count is said to be quick and slow, nor that which is counted. For how could the latter have a quick and slow if that with which we count has not? For since it is a number on account of the former, so if the number by which we count were quick and slow, so the numbered would receive the quick and slow
731,1 from it. It is also clear that magnitude *qua* magnitude has no quick and slow, for these are properties of change, not of magnitude. He also showed previously[196] that time has no quick and slow, but change has. For this is determined by time. But time is not determined by time.

5 > **220b5-12** Also there is the same time everywhere simultaneously; [but the time before and after is not the same. For the

[195] Simplicius writes *takhus kai bradus oudeis esti*. Ross omits *esti*, nor does he mention it in his apparatus. (J.O.U.)
[196] Aristotle *Physics* 218b14ff. (J.O.U.)

present change is one, but the change that has occurred and that impending are different. Time is a number, not that by which we number but as what is numbered, and this comes about before and after, always different, for the nows are different. The number of a hundred horses and of a hundred men is the same, but the things of which it is the number are different] – the horses from the men.

One of the things previously said about time was that the present, viewed at a now, was everywhere simultaneously the same, but the before and after were not the same. That this, like what immediately follows, accords with the received concept of time he proves from the fact that change, in which time has its being, has these features. For the present and occurring time is the same, since it has no internal difference, but is one through the common form of the extension of its being. But a past and future change are different from each other. So as one change exhibits identity and difference, so with every time which simultaneously measures all changes. For time is not the number of this and that change as such, but of change *qua* change. So there is one time of the changes that are many, since it is not their number *qua* many, but *qua* all being changes. But the present change is all one *qua* change.

Time is not merely one in form, but is so of a change that is one both in form and numerically, since nothing in it contains differences. But the time before is other than the time after, since the change also, even if taken as one in number, is other as before and after, since it has its being in becoming.

He next demonstrates the point by the fact that the time before and the time after are not the same. For the number by which we number is always the same, whether or not it contains a before and after like a triad, but the things measured must necessarily be other; for horses are different from men. So if time is a number as being that which is numbered in change, and not as that which numbers, but what is numbered in change does not remain the same, but is taken as other, it is to be expected that the time before will be other than the time after, and that which is past from that to come. For the nows also by which it is numbered and determined are other. For the numbered in change are different[197] as before and after, and clearly different in account, though the same in substrate and form. For this was demonstrated before.

220b12-14 Further, as it is possible for one and the same 732,1

[197] Either Simplicius has made a grammatical slip or *to gar arithmoumenon* in Diels's text should be plural. (J.O.U.)

> change to recur again and again, [so for a time,] such as a year or a spring or an autumn.

He has said that the present occurring change is all the same in its universality and that therefore the time that is present is the same everywhere, but the past and the future change is other, and the times other also; he now shows the way in which also the before and after both in change and in time can be the same through recurrence. Having said before that the now is identical with the substrate, though different in account, I think that he now articulates how he was then using 'identical', and that it was in the way that a change can be so by recurrence, i.e. by happening again and again in the same form. For since changes recur, as the sun goes again and again from the same place, again and again from the Ram and from the Plough,[198] so time accompanies change. It is clear that a year and a spring constantly recurs, as does an autumn.

But why, then, did he neither say earlier that the now was the same in form but that it was the same in substrate, nor say now that a time was the same in form, as he said of change, but by recurrence? Yet he especially insists on the threefold exhibition of the same thing by genus, form and number. Perhaps it was because not only do things in the same genus allow much difference between them, but also even things said to be the same in form still contain many inessential differences, if they differ numerically as do Plato and Socrates. But changes, at any rate, which occur over and over again the same, and the nows and the times, are almost numerically identical.

Thus the Pythagoreans also said that things became in a way even numerically identical by recurrence. Perhaps it would be no bad thing to hear Eudemus' discussion from the third book of his *Physics* where he paraphrases their views: 'One might raise the problem whether the same time recurs, as some say, or not. "The same" has many senses: the same in form seems to occur as do spring and winter and the other seasons and periods; similarly the same changes occur in form, for the sun performs its solstices and equinoxes and its other journeys. But if someone were to believe the Pythagoreans that numerically the same things recur, then I also will romance, holding my staff, while you sit there, and everything else will be the same, and it is plausible to say that the time will be the same. For of one and the same change, and similarly many other identical things, the before and after will be the same, as will their number. Everything will be the same, and so time also.'

733,1 Observe that he calls the change that recurs one and the same, and the before and after one and the same. But to see that Eudemus

[198] Constellations.

also calls what continually happens one thing after another in the case of both change and time, listen to what he wrote a little after the above quotation: 'As in a change an atom exists only, the past has gone and what is to be is not yet, so one part of time has departed, another is not yet, and what is present is always one after another.'

> **220b14-32** We do not only measure change by time, [but also time by change, since they are determined by each other. For time determines change by being its number, and the change time. Also we call a time much or little measuring it by change, just as we measure the number by the thing numbered; for example, the number of horses by the unit horse. For we know the quantity of horses by number, and again the actual number of horses by the unit horse. It is the same with time and change, for we measure change by time and time by change. This happens as one might expect; for change is in accordance with magnitude, and time with change, through their being quantities and continuous and divisible. For because the magnitude is so great the change has such and such a character, and time because of the change. Also we measure magnitude by change and change by magnitude; for we say that the road is long if the journey is long, and call that long if the road is long. And we call the time long if the change is,] and the change if the time is.

Since time is called the number and measure of change, since change is measured by it, and a change is counted as quicker or slower as it is measured by less and more time, he now adds that not only does time measure change, but also change time. Each is measured by the other through their being determined by each other, change by time and time by change. For a time in which change is great, is itself counted as great. Pre-eminently time measures change as being its number and determining it by the before and after, but in a contingent way it is measured by it. We see the same thing happening in the case of all other measures. For the measures are measured in their turn by the things measured. For we judge the wooden bushel-measure by the bushel of wheat, whether it is not too small or too big, and we measure the bronze pint-cup in its turn by the pint of wine.

Having said of time and change and the other continuous measures that they are measured reciprocally by the things they measure, he says that one may see the same thing in the case of numbers. For number is in a certain way numbered by the things it numbers. For as we speak of ten horses, because we have numbered them by ten units, so we say that there are ten units because they equalled ten horses. For perhaps we would never have gained the concept of ten if we had not taken that there was one horse and one man so many times.

Having said this, he produces the explanation of the reciprocal measurement. 'For change', he says (see above), 'is in accordance with magnitude, and time with change.' So as change is to magnitude, so is time to change, and change measures magnitude and is measured by it in turn. 'For we say that the road is long if the journey is long' and that the journey[199] is long if the road is long. Time must have the same relation to change. For just as a change is great if the time is great, so a time is great if the change is great. For being equalled to each other they measure each other. Also, time being continuous and divisible on account of change, what is remarkable if it is also measured by it, just as change also has these characteristics through magnitude and is so measured by it?

Different readings are recorded at the end of the passage above. If it is as follows: 'and [we call] the time [long] if the change is, and the change if the time is', as in most manuscripts, what is said is clear. But what of 'and [we call] the time [long] if the change is, and the time the change'? This reading is recorded and Alexander rejects it, saying that the two clauses mean the same, that time is measured by change. But how could 'and time the change' mean that time is measured by change? But having said 'and the time if the change is' to confirm that change measures time, according to this conception he vigorously added 'and the time the change'.[200] It is worth noting that he now still says that the change is measured by magnitude and time by change. But the before and after allotted to the change from its position in the magnitude is different from that according to how long it continues, according to which time is viewed. So one must also examine the ratio which says that the extent of change is as that of magnitude, and the extent of time as that of change, to see how it can be correct. For if the road and the journey are both great, nothing prevents the time being short, if the movement is quick. Or it is possible, considering these matters for ourselves [or: in themselves], that what would be much time in the case of change is the time of much change. Or it is possible to take the much and the little of time analogously to speed; for, for the fastest runner, one

[199] Simplicius writes *kai pollên tên poreian*. Ross's text of Aristotle reads *kai tautên pollên*. I substitute a full stop for Diels's comma after *pollê* at 734,2. (J.O.U.)

[200] The translation fails to bring out the problems, one being that in the second reading the word 'time' in the clause 'and time the change', cannot be the object of 'call' since it is in the nominative. Presumably we must very awkwardly supply some such word as 'measures' so that the clause expands to 'and time measures the change'. (J.O.U.)

day would be a long time to take to run the long road from Elis to Epidaurus.

> **220b32-221a9** Since time is the measure of change [and being changed, and since it measures the change by determining a certain change that will measure the whole, as the yard does the length by determining a certain length that will measure up the whole: and since also for change to be in time is for both it and its being to be measured by time – for time measures at once both the change and its being – both this is what it is for it to be in time, that its being should be measured, and it is clear for other things also that this is what it is to be in time], that their being should be measured by time.

He has shown that time exists, and that its accepted characteristics accord with the definition, and that we primarily say things are in time when they are contained by time. Now, I think, he makes the 735,1
most essential points about time, and shows why each thing is said to be in time. But the long wait for the apodosis in the exposition[201] has caused a certain unclarity. For the apodosis of 'since time is the measure of change and being changed' is that for change and for other things to be in time is 'that their being should be measured by time'. The conjunction 'both' in 'both this is what' is not, I think, superfluous, as Alexander thought, since he joined the clause to the preceding [by taking *kai* to mean 'and']. But perhaps, as Alexander also judged, the apodosis is from 'it is clear that for others also this is what it is', 'others' being taken to refer to change and not to all things. For Aristotle's concept is that by the medium of change for all things 'this is what it is to be in time, that their being should be measured by time'. This is reasonable, for the being of each thing is the activity and change of its being, as if he had said 'since time is the measure of change both for change and for other things in regard to their change, this is what it is for them to be in time – for them to have their being measured by time.

He demonstrates that change is in time as follows: time measures change, but to measure change, which does not exist all at once but has its being in becoming, is properly the same thing as to measure its being, that is, the quantity of its existence. For a change is measured also with regard to the interval through which the change

[201] The apodosis is the clause beginning 'both this is what' in the translation of Aristotle's text above. This requires a punctuation different from that of Ross. But Simplicius will also consider a reading more like that of Ross, in which the apodosis begins at 'it is clear that', in which case the translation should end 'for time measures at once both the change and its being, and this is what it is for it to be in time, that its being should be measured – it is therefore clear that for others also this is what it is to be in time, that their being should be measured by time'. (J.O.U.)

occurs, as when we say something moved a furlong. But the interval is measured incidentally and not *qua* change. For this sort of measure requires that what was before remain; but as a change and as having its being in becoming, its measure is how much time it takes. For this is the measure of its being as having its being in becoming.

See now how clearly, as I think, Aristotle has expounded how time is the measure of change, that it is through the extension of its existence, which is principally how it has its being. 'For', as Alexander says, 'in the case of change, it is the same for it to exist and for it to be a change, as for all other things that have their being in becoming. For that reason it is the same thing to measure the change and to measure its being.' It is clear that this is another sense of 'being' from that customarily used by Peripatetics, which means 'form'. For this is clearly different and refers to the extent of its being and, as it were, the actuality of its existence. Since, then, the being of change is an extended actuality – for it is the realisation of the changeable – it is to be expected that change and the existence of change should be the same. That is why Aristotle himself, having said that time was the measure of change, added 'and of being changed', not as something additional but because he has shown that change is an activity, and that time is a measure of activity. 736,1 That change and its existence are measured by time he shows by the fact that all change that has elapsed is determined by time. For future change does not yet exist and therefore has not been determined.

Just as he who says that time measures the existence of man says so because time determines and measures the number of years a man lives, so, in the case of change, he who says that the existence of a change is measured by time says nothing other than that time measures the extent of the change. In the case also of other things that do not have their being in becoming but exist complete, as in the case of a man, time measures out the existence of the man, that is, the period of his being; but the man himself, that is the interval he occupies, is measured in feet or something similar. But in the case of change time measures both the existence of the change and the change itself. For, in the case of change, the existence of the change that is measurable by time and the change itself coincide.

But, if this is so, what is it in a change that is measured by magnitude, if we not only measure the magnitude by the change but also the change by the magnitude, as has been said? Is it not clear that it is its extension, which is the same as the substrate by the extent of its existence; for it is an activity and does not endure. It is measured as an enduring interval with reference to its magnitude by magnitude (for we say that a motion was a furlong), but as the extent of its existence it is measured by time. But this being the same as the substrate, change and its existence are reasonably said to be the same thing.

Having said that time was the measure of change, he adds how this is done. For everything measured seems to be measured out by a portion of itself, since that which measures the whole is a part of it, so that it contains the number of measures that go into it. So it will be like that in the case of change also, if it is measured by change. For a part of a change is a change, and every continuous thing can be divided into equal determinate quantities. How then is a change said to be measured by time? Time, he says, marks off a portion of a change, by which the whole is measured. The prior and later nows mark it off. Time measures change in much the same way as a yard measures magnitude, not as a portion of it, but by marking off a portion of it. The whole is measured by the portion being taken so many times. Since then the change marked off by time, such as one lasting an hour, measures the whole of the change, the change is therefore said to be measured by time.

Having shown that that is what it is for change to be in time itself, or, in other words, for its existence to be measured by time, after change he turns also to the other things that are said to be in time,
just as if he had said that if (1) for change to be in time is for its 737,1
existence to be measured by time, so (2) for other things also this is what it will be for them to be in time – that their existence should be measured by time, because their existence is the activity and change of their existence. But (1), therefore (2).[202] But he did not go so far as to say that, in the case of other things, both they and their existence were measured by time, because, in the case of change, change and its existence were the same, but, in the case of other things said to be in time they are not the same. The interval of their existence had its being in becoming, but not they themselves. So they are in time because their existence also is measured by time.

I think that difficulties arise from this. First, if time is the number and measure of change, how is anything else beyond change measured by time and in time, so that one can say that for other things also this is what it is to be in time? For things other than change, not being change, are also not measured by time, if time is the measure of change.[203]

Secondly, how could this man and this horse and each of the other

[202] This is a Stoic mode of presentation for arguments.

[203] Diels has no question-mark after 'time' in 737,14, but a colon. His next sentence reads *ta gar alla tês kinêseôs, ou kinêsei de ê kinêsis oude hupo khronou metreitai, eiper kinêseôs metron estin ho khronos*. I have no doubt that my text conveys the correct general sense of the difficulty raised, but it can hardly include a correct translation of *ou kinêsei de ê kinêsis*, which I do not understand. (J.O.U.)

things which do not have their being in becoming, in the way in which it is possible to say this of things in the world of becoming, be in time, since time has its being in becoming, and it is necessary that what is said to be in time should be correlated with time?

Perhaps one should say in reply to the first problem that, since his discussion is framed primarily for the case of change of place, as is shown by his saying that change is measured by magnitude, and that this is great if the road is great, by 'the other things' he means the remaining kinds of motion (*kinêsis*) and change (*metabolê*) – growth, alteration and coming to be. But it is better, perhaps, to say that change is an activity and has its being in becoming, which is the same thing as change. The being of change is pre-eminently in time and is measured by time, while the other things that are said to be in time, such as a man or a horse, are themselves so through their essential change, which means that their being is an activity of their essence and being, an activity that is neither complete nor permanent, but has its being in becoming. It is in this way in time, as a compound of coming to be and passing away. So taking the other things that are said to be in time in addition to change, he says that they are in time through their internal change, which is the extension of their being. So the things that truly are and do not have their being in becoming, which both Plato and Aristotle call eternal, are not in time but in eternity that is statically in unity.[204]

If this is correct, the second problem is also solved. For generated substances are said to be temporal because of their internal flow and extension. So, time is the measure of the flow and extension of existence. They are said to last as long themselves as the time that their existence extends. This, as it seems to me, is the penetrating and apposite philosophical teaching of Aristotle about time.

> **221a9-18** For to be in time is one of two things. [One is to exist when time exists, the other as when we say of some things that they are in the numerical system. The latter can mean either that it is a part of number and an attribute, or that it has a number. Since time is a number, the now and the before and the like are in time as a unit and the odd and the even are in the numerical system, the latter being elements in number, the former in time. But things are in time as a number. If this is so, they are contained by time] as things in place by place.[205]

[204] Simplicius writes that these things are *aïdia* and *en aiôni*. This is surprising, since in Aristotle's terminology, and often in Plato's, the *aïdia* are the everlasting, such as the heavens, while the *aiônia* are timelessly eternal and *en aiôni*. But Proclus says that *aïdios* is generic and includes the *aiônios* (*E.T.* 55) and Simplicius is apparently following that usage here. But see 739,13-20 below. (J.O.U.)

[205] Simplicius writes *hoion ta en tôi topôi hupo tou topou*. Ross differs. (J.O.U.)

Having said that change is said to be in time as being measured by time, he next enumerates the senses of 'being in time', to make it clear how temporal things such as the now, the before and the after, and the like are in time, and how other things are so. For 'in time' means two things: one is to exist when time exists, one as we say that things are in the numerical system.[206] This, too, has two senses. Things may be in the numerical system as a part or an attribute of number, as two and the odd, one being a part, the other an attribute. Or it may be that they have a number and are enumerated, like ten horses and ten oxen are in the numerical system.

So, since time is a number, the state of things in time will be the same as of those in number. So the now and the before are in time in the same way as the unit and the odd are in the system of abstract numbers. For they are parts of time and of number, respectively. But change and the other things that are not parts or attributes of time are in time in the way that ten men are in number, by being numbered by it. Just as for things that are in number in this way there is always a greater number, so for things that are in time there is always a greater time. For time also is a number. If this is so, things in time are contained in time in the way that things in number are contained in number and things in place are contained by place and everything else that is in something as a container.

Notice how Aristotle recognised the affinity of time to place and number before those who pointed it out later. This is more important than the signification of being in time. For to say that something has its being in time when there is time is a less precise account of being in time. In that way the unchanging could be said to be in change, if it is when there is change. How does this differ from the universe being in a grain,[207] i.e. generally the whole in everything, and the greatest in the smallest, if the grain exists when
the universe exists?[208] But many things happen to exist 739,1
simultaneously, which is what 'then when ...' means. But that one should be in the other is for the one to be contained in the other. It follows in this case that the things are simultaneous, but they are not in each other because they are simultaneous; they are simultaneous because they are in each other. And this is because both are in time. But things that are not in time, even if they exist when there is time would not be said to be simultaneous, since those things are simultaneous that are in the same time. Nor is it proper

[206] The Greek is *en arithmôi* – in number. But we do not use the expression as the text requires. I follow Ross in using the expression 'in the numerical system'. (J.O.U.)

[207] Cited above as the absurd consequence of allowing bodies to be in the same place, 530,23-4.

[208] Diels has a full stop instead of a question mark. (J.O.U.)

to speak of them as being 'then when'.

Having shown that things are properly said to be in time as being contained in time, as things in number by number, he produces a sign of things being thus in time and at the same time an explanation why this is so. For in the case of things properly in time it is always possible to find a time greater than their existence. In that way they will be contained in time and properly said to be in time, as in the cases of place and number. 'That', says Alexander, 'is why everlasting things are not in time; for time does not contain their existence.' If by 'everlasting' he means 'eternal', as Plato did in the *Timaeus*[209] in regard to the paradigm,[210] saying 'the nature of the animal happened to be everlasting', he would be right. For the eternal is not in time nor is contained by time. But if by 'everlasting' he means that which exists throughout all time, it is not true that this is not in time, if it is required only that its existence be in time. For such things have their being in becoming and are not simultaneous wholes like eternal things. Alexander himself bears witness to this in the passage where he writes: 'Change, being everlasting, is in time, since it does not exist as a substance nor remain numerically identical, but has its being in becoming. It for ever becomes something different, and is thus in time.' If, however, the other things also that are in becoming through change are in time – for that is their nature – if some of them have an inexhaustible existence consisting in becoming, changing either in activity or in quality or in place, there is nothing that stops such things being in time. But to be in time is like being numerable in every respect, since time is also a number. For of things that are numerable it is possible to find a greater number, since everything enumerated is finite in number. But it is possible to find a number greater than any finite number, since number increases *ad infinitum*, just as magnitude decreases *ad infinitum*. But it is not always possible to find a place greater than occupied place. For, if place is the limit of the container,[211] or the ordering of discrete
740,1 things, place is circumscribed also by place; but, if place is the interval, individual bodies in place always have some place exceeding them, into which they often change place. But the whole no longer admits a greater place. For the whole includes the whole of place. So of everything called place it is true to say that there is not a

[209] Plato *Timaeus* 37D. Here 'eternal' translates *aiônios* and 'everlasting' *aïdios*. (J.O.U.)

[210] The Platonic Form of the living universe.

[211] The first definition is Aristotle's in *Physics* 4: my place is the inner surface of what surrounds me. The second is favoured by Simplicius in his Corollary on Place (translated Urmson): my place is dynamic and keeps my limbs correctly assembled. The third is favoured by Philoponus in his Corollary on Place (translated Furley), but Philoponus seeks to avoid the consequence that the interval would need to be infinite.

place greater than every place, as there is of every time and every number. But he used the illustration of place only to demonstrate that things in time are contained in time, as things in place by place. Alexander summarised the whole account as follows: things in time are in number; but there is a number of things in number. Therefore there is a number of things in time. Of things that have a number there can be a greater number: for number is finite. But it is possible to find things exceeding anything finite. So of things that are in time there is a greater number in respect of time.

> **221a30-221b3** Things are also affected by time, [as we are accustomed to say that time wears things out, and that everything ages through time, and is forgotten because of time, but not that someone has so learnt or become young or beautiful. For time is rather a cause of decay in itself. For it is the number of change] and change ends that which is.

He has shown that what is in time is contained by time, and contained in such a way that there is time before it and after it. He now adds that things in time are affected by time; 'for we are accustomed to say that time wears things out, and that everything ages through time, and is forgotten because of time', and these are as such destructive of life and knowledge. For we do not similarly say that someone has learnt or become young or beautiful because of time, since time seems to be the cause of decay, being the number of change. But change is departure from being and is a change. So it is not surprising that time, being related to change, ends the existence of that which is in it and causes it to change, since it is the number of the change and cessation of everything that is in it. He properly added 'rather' in saying 'is a cause of decay in itself', because he does not think that time is absolutely and essentially the cause of things being affected, but only in so far as everything in time is subject to being affected. Also it is clear that it seemed to be the cause not only of decay but also of coming to be and emergence to him who said

> long and uncountable time brings all about;
> it breeds the unseen and hides the seen.[212]

And it is the cause not only of forgetting, but also of learning, as we may learn from him when he said

> time gives me some knowledge.

[212] Sophocles *Ajax* 646-7. (J.O.U.)

741,1 Simonides[213] attributed to time the greatest wisdom; for he said that by it all men make discoveries and learn.[214] … But Euenus[215] combined the two into

> time that is wisest and never learns.

Further, Aristotle himself said in Book 3 that change is of two types, one from form to privation and one from privation to form; so change does not put an end only to form, but also to privation. The reason why it is rather the cause of decay is that, when things come to be, even if they seem to do so in time, it is not because of time, it being clear that the efficient cause is something other than time, as the builder is of the house. But things that decay do not seem merely to decay in time, but also because of time, since nothing else appears to be the cause of the decay. Similarly the teacher causes learning, but time seems sufficient for forgetting.

> **221b3-7** So it is clear that things that exist for ever, *qua* existing for ever, [are not in time. For they are not contained by time, nor is their existence measured by time. A sign of this is that they are unaffected by time] through not being in time.

Having shown that those things are in time for which it is possible to find a longer time, and that things in time are affected by time, he thence deduces as a conclusion that 'things that exist for ever, *qua* existing for ever, are not in time', for they are neither contained by time, through it being possible to find a greater time than their existence, nor affected by time. He added '*qua* existing for ever' because, though they themselves are co-present with time, their motions (*kinêseis*) and changes (*metabolai*) are in time because it is possible to find a greater time than each of them takes.

What then of everlasting change? For it will be shown that the circular motion is everlasting. Is that in time, or is it not? If (1) it is not in time, time is not the number of all change; if (2) it is in time, how is that in time which no time exceeds? Is it because, since it is always different and never numerically identical, it is always possible to find a greater time than the motion taken? Themistius says: 'Even if time does not exceed this change, nor contain it in that way, still, since they are interlocked and co-extended and becoming other together, the change is therefore in time.' But perhaps
742,1 someone might say that on that account time was in change.

[213] Poet, 6th-5th century BC.

[214] There appears to be a lacuna. Diels suggests supplying Pythagoras' statement that time never learns, referred to at 745,15. (J.O.U.)

[215] Poet, 5th-4th century BC.

But of things called essentially everlasting some for ever exist as simultaneous wholes, such as things intelligible and truly real, and are always fully active, and are entirely beyond time, being measured by eternity. But the celestial being[216] is very clearly essentially unchangeable; if, however, it suffers no alteration it is clear that, having in its being no change nor extension, it also will be beyond time. For time is the measure of things that are coming to be, not of things that are. But if each implants in the other a certain completion in accordance with their different arrangements, as we see plainly happening in the case of the moon, which differs at different times and completes her being, it is clear that in those respects the celestial will be measured by time. For what participates in becoming must in some respect be coming to be and passing away, but not necessarily in essence. For Aristotle and his followers who always philosophised from the obvious, it suffices that the essence of heavenly things should be ungenerated and beyond time, but their changes should be temporal. Perhaps it is better to say that at the extremes some things are wholly unchanging and some things wholly changing, and that they are measured by eternity and time respectively, but in between some things like the soul and everlasting bodies are measured in intermediate anonymous ways.

221b7-20 Since time is the measure of change, [it will also contingently be the measure of rest. For all rest is in time. For it is not necessary for everything in time to be changing, as it is for things in change: for time is not change, but the number of change, and it is possible for something at rest to be in the number of change. For not everything that is unchanging can rest, but only that which is devoid of change but naturally changeable, as has been said earlier. But to be in number is for there to be some number of the matter, and for its being to be measured by the number in which it is, so, if it is in time, by time. But time will measure the changing and the resting, both as change and as rest. For it will measure both the change and the rest of things, being a certain quantity. So that which changes will not be measured as such,] but in so far as the change is a quantity.

Having, so far, said that time is the measure of change, he reasonably adds that time is also the measure of rest, and is so because it is the measure of change. This is shown by the conjunction 'since'. For if rest is the privation of change, and if the

[216] The heavens.

same things distinguish and measure both possession and privation, as sight is of light and darkness and hearing of sound and silence and the ruler of the straight and the crooked, one would expect that that which measures change will also measure its privation, which is rest. But this it will measure contingently. For essentially time measures change, for its being consists in so doing, but contingently and by reference to change it measures rest also. For in rest it measures the absence of change, as he himself will say. For there is a certain time for which that which rests is at rest, than which one can find a greater. For those things are said to be at rest that naturally change.

Alexander says that time is the number of rest because privation of change is an attribute of change, of which as such time is the number. But Themistius says that it is because it measures another change: 'For night is as such the measure of the motion of the sun under the earth, but contingently of the rest of living things.' But one could put the matter otherwise. For since every rest is between two changes which time measures as such, and the earlier now is the limit of the earlier change, the later now the beginning of the second, the interval between would be the time of the interval of intermediate rest contingently, because there happened to be rest between two changes. But how does time, whose essence is to be the number of and the measure of change measure rest also, which is the privation of change? For rest also will be a change if what is measured by time is change. Answering this objection, he says that it is not necessary for what is in time to be changing; 'for time is not change, but the number of change' and an attribute of it. If something is in an attribute it does not have to be in that of which it is an attribute. What holds of a day does not have to hold of the motion of the sun, just because a day is so much time that is the number of so much motion of the sun. So rest does not have to be in change, and there is thus no obstacle to time being the number of change. For we say that a sick man was at rest in so much time. Moreover, the ruler that is as such the criterion of straight lines incidentally detects crooked ones as well.

Having said that what is at rest is in time, he says that not everything that is unchanging is at rest. If it were, time would not be the measure of all rest. For it is not of the eternal unchangingness, since it does not contain its being.[217] Those things are at rest, as has been said, that, while naturally changeable, are deprived of change. But what is naturally changeable does not have everlasting changelessness. He then adds what it is to be in time,

[217] Simplicius says at 23,13 that the word for eternal exemption from change is *monê*, not *êremia*, here translated as 'rest'. (J.O.U.)

that, just as to be in number is to be some number of the matter and for its existence to be measured by the number in which it is, so to be in time is for its existence to be measured by time. To which he adds that time will measure that which is at rest as such, just as it measures what is changing as such. But it does not measure the substrate, but only their change and rest. ‘So’, says Alexander, ‘time will measure the changes of everlasting things also, but not the being of the things that change.’ Some people say on the basis of what is said here that the definition of time is incomplete; for they say that it ought to have been defined as the number not only of change but of rest also. Criticising them Alexander says: ‘They so understand “the number of change” as that it is the same as “the measure of change”. So they think that, since it is the measure of 744,1 rest also, and not only of change, this should be added to the account. But they are not the same. The being of time does not consist in its measuring change; that is an attribute of it. Its being and essence lie in the measurement of the before and after of change *qua* before and after, which is not so of rest. For rest does not consist in a change from something earlier to a second after it, as does change. Aristotle made this clear when he said that what is at rest can also be in the number of change; but to be in a number is to be some number of the matter.’ When Alexander says this I am led to wonder what he means when he says that rest does not have a before and after, as does change. If it did not have it, but the earlier now was fused to the later, there would be no awareness of time during rest, as there would not be of change. But if the earlier and later is a change from this[218] to that, and, the earlier passing away, the later takes its place, why does rest not have its being in becoming also, so that some change will appear in it? So perhaps there are two kinds of change, one which is opposed to rest and one also that is the extension of the existence of rest. For, as we previously called the flow of the presence of everything else in becoming, their existence, so the being of rest consists in flow and extension. That is why Aristotle also, though he first said that rest was contingently measured by time, goes on to say that ‘time will measure both what is changing and what is at rest, the one *qua* changing, the other *qua* at rest – both change and rest, as being some quantity’ – i.e. it will measure that duration of each, in their existence, which is their coming to be and their flow. Perhaps this is why he said that rest was measured contingently by time – because it is not as rest as such but in respect of its generated existence, just as change and other kinds of thing are timed in that way. The

[218] The text has *ekeinês*. Diels writes ‘suspicor *ekeinou*’, which I have adopted. (J.O.U.)

difference is that, since in the case of change it and its existence are one and the same, time seems to be essentially present in change.

It is clear that he does not say that rest is the privation of change as the crooked is of the straight and generally the unnatural of the natural, but as being the absence of a natural state while itself being natural, since he defines nature as the principle and cause not only of change but also essentially of rest. For nature is the cause of rest also. Perhaps the reason why he did not call it being stationary is that he wished to have a special term for the stationary state from which and into which change changes. For even things that remain for ever unchanging like the poles and the axis of the universe are said to be stationary, but not to be at rest, since 'at rest' applies to things that sometimes change.

> **221b20-3** So those things that neither change nor are at rest are not in time. [For to be in time is to be measured by time] and time is the measure of change and rest.

Having earlier said which things are in time, being those things that change and are at rest, and which sometimes exist and sometimes do not and are outlasted by time, in this passage he shows what things are not in time, because 'those things that neither change nor are at rest are not in time', and, in the next passage, that the totally non-existent and the eternal are also not in time but only those things that come to be and pass away. He shows that things that neither change nor are at rest are not in time as follows: if time is the measure of change and rest, those things that neither change nor are at rest are not in time. The things that neither change nor are at rest are those that are unchangeable in their own nature and, obviously, things that undergo no form of change. The syllogism is in the second figure[219] as follows: Things in time are measured by time; things that neither change nor are at rest are not measured by time; conclusion: things neither changing nor at rest are not in time. He set out the major premiss by 'for to be in time is to be measured by time'; he did not set out the minor premiss that says that things that neither change nor are at rest are not measured by time, but that which supports it, which says 'time is the measure of change and rest', where 'only' has to be understood as already established. If it is the measure of these alone, it is clearly true that that which neither changes nor is at rest is not in time. Since of things that exist some are unchangeable, some for ever changing, some sometimes changing, sometimes at rest, it is clear that the altogether unchanging is not in time, but things that are always

[219] The second figure is of the form PM, SM, therefore SP. (J.O.U.)

changing and those that sometimes are and sometimes are at rest are in time. Things that change neither always nor sometimes are not in time. The things that sometimes change are clearly those that are sometimes at rest.

But how is time the measure of change and rest? For those who 746,1
say that time is essentially the measure of change, but of rest as the privation of change, naturally say that it is the measure of these alone. But those who say that time is the measure only of change, and of rest only as being also in flux, time will clearly be the measure of the existence not only of rest but of all other things that are in becoming, as Aristotle said himself.[220] Time is primarily the measure of change in the flux of existence, which is the property of becoming. But perhaps, since everything in becoming is either changing or at rest, it is true to say that these alone are in time.

> **221b23-222a9** So it is clear that that which does not also exist will not all be in time, [such as things which cannot be otherwise, like the diagonal being commensurate with the side. For, in general, if time is essentially the measure of change and of other things contingently, it is clear that the things whose existence it measures will be all those whose being is in change and rest. So all things that perish and are generated, and in general sometimes exist and sometimes do not, will necessarily be in time. For there will be a greater time which will exceed their existence and the time that measures it. But of non-existent things which time contains some used to exist, as Homer did, some will exist, as anything in the future, depending on from which side it contains them. If it is from both sides, they will be both past and future. Things that time in no way contains neither were nor are nor will be. These are the non-existent things whose opposites are everlasting, like the incommensurability of the diagonal, and that will not be in time. Nor will its commensurability be, so it is for ever non-existent as the opposite of the everlasting. Things whose opposite is not everlasting may either exist or not,] and they come to be and pass away.

He still continues to articulate which things are in time or not. He now shows that neither things altogether non-existent nor things everlasting, nor in general things not exceeded by time, are in time, but only those things which come to be and perish. In this way he makes clearer the conception that he has of time. He says that not all non-existent things are in time, but some. For of the non-existent

[220] I reproduce the anacoluthon in the Greek. The sense is clear. (J.O.U.)

some have passed away, like Homer the poet, some are future like an eclipse, some are both past and future like a war and an eclipse.[221] All these will be shown to be in time. But some non-existent things cannot exist, as a diagonal being commensurate with the side cannot possibly exist. So such things are not in time. He adds the explanation. It is that time is essentially the measure of change and of other things contingently. By 'other things' he means rest itself and those things themselves that change and are at rest. Of these time measures rest itself, also 'measuring', says Alexander, 'the extent of the absence of change in what is at rest'; but I think also the duration of the rest. He says 'contingently' because of the dependence on change. But time does not initially measure the substances themselves that change and are at rest, but it is said to measure them contingently because the things measured by time, the change and rest, happen to the changing or resting thing that is a man or a horse. Since, then, time measures only change and rest, clearly those things whose existence is measured will have their being in changing and resting. These things are those generated and perishing and, in general, whatever sometimes exist and sometimes do not. That is why such things are contained by time. For there is a greater time that exceeds that which measures their existence. So of non-existent things such as once existed or will be or both existed and will be, of these time exceeds the time of their non-existence; for there is the time in which they were or will be. That is why such non-existing things are in time.

Having given an example of the impending he added 'from which side it contains them'. For, in the case of what used to exist, their non-existence, or the time that measured it, is bounded by the past time in which they did exist; so such non-existent things are in time. In the case of impending things, their non-existence is bounded by future time. For the time in which they exist together with that in which they do not is greater than that in which they do not. So those things are also in time. Those things that were before and will be later, but are not now, like an eclipse, an omen or a war are such that the time of their non-existence is contained from both directions – both by past time, since they existed in it, and by future time, since they will exist in it. Also each time is greater than that alone in which it does not exist, both that including the past and that including the future. So in the case of these things also, their non-existence is in time, as is their existence. For a time exceeds each of them, and therefore they are contained in time and in time.

[221] It is tempting to replace 'an eclipse' on its first occurrence by 'the flood' (*ho kataklusmos*) as an example of the future. See 750,17 for the flood as an example of the future, of which an eclipse is a poor example, since eclipses, like wars, recur. See 747,11. (J.O.U.)

But the non-existence of things which cannot be is not in time. He first said which things were by means of an example – 'like the diagonal's being incommensurable with the side', but now laid down a universal rule. It is impossible for those things to exist whose opposites are everlasting.[222] Going back, he gave a general rule for which things can both exist and not exist, saying that they are such whose opposites do not exist for ever, but, clearly, sometimes do. For, if their opposites always existed, they would never do so; but if their opposites never existed they would always exist. But if their opposites sometimes exist and sometimes do not, then they will sometimes exist and sometimes not, coming to be and passing away. When their opposites exist, they do not; when those do not exist, they do.

CHAPTER 13

> **222a10-20** The now holds time together, [as has been said. For it holds together past and future time, and it is the limit of a time. For it is the beginning of one and the end of another. But this is not plain as in the case of the point that remains. It divides only potentially. In so far as it does so, the now is always other, but, as joining, it is always the same, just as in the case of mathematical lines, where the point is not always the same for thought, being different for those who are dividing, but as one totally the same; similarly the now is both potentially the division of time, but an end point of both and a unity. Division and unification are one and the same thing and in the same respect, but their essence is not the same].[223]

He proposes to speak about the now,[224] in which time seems most of all to consist, both in the strict and in the colloquial sense, and first of all about time in the strict sense, of which he has previously said that time is continuous through the now and divided at the now. He 748,1
will now say the same about it at greater length, that the now binds time together. That is so of time *qua* continuous. For those things are continuous whose parts meet at some common boundary, as will

[222] Here, as elsewhere in this discussion, I have translated *aïdios* as 'everlasting'. Strictly, Aristotle, and Simplicius following him, would have said that such things as mathematical truths were *aiônia*, eternal. (J.O.U.)

[223] Diels's text includes in the lemma the words *to men oun houtô legetai tôn nun*, which are the first phrase of a new sentence. They are given in Simplicius as the beginning of a new lemma at 749,33. Manuscript a excludes these words, and I follow it. (J.O.U.)

[224] The present instant, and hence derivatively any instant.

be shown in Book 6 of this work.[225] The parts of time, the past and the future, are joined together at the now, which becomes a common boundary belonging to both. So considered, the now holds time together. But when it is taken and conceived as the limit of the past and as the beginning of the future, it now divides it. For the same thing sometimes holds together, sometimes divides, as is the case with a point in a length. For it, too, both holds the length together, when it is conceived of and treated as one, and it divides, when we use it as two, calling it the limit of one and the beginning of the other. That is what Alexander says, thinking that continuity is being explained as far as 'for it holds together past and future time', but understanding what follows as being about division, because Aristotle called the now the limit of one time and the beginning of the other. But it does not hold together as remaining as in the case of the point. For the now of flowing time does not remain.

The manuscripts differ, one having 'and a boundary of time' (*kai horos khronou*), the other 'and a limit of time' (*kai peras khronou*). The sense is the same in both for, continuing, he calls the common boundary the limit of both and their unity, but 'boundary' is more apposite, since we call those things continuous that are joined together at a common boundary.

Also Alexander takes 'and divides potentially' as going with the preceding statement that it is not plain as in the case of the point that remains, while the now divides potentially. Such an interpretation is reasonable; so he says that the now divides time potentially, that is, for thought. For it does not do so actually, like a point, by having a position and remaining and exhibiting the parts as separate; for, because of the flow of time, the parts of time cannot be exhibited separate from each other.

When the now is taken as holding together it is the same in account, and not only in substrate. For it does not hold time together when taken as having different accounts, but when it has the same account with regard to both parts, becoming the common boundary of both. It is, he says, the same in the case of the now as in that of mathematical entities.[226] In so far as (1) we take what is in substrate the same point as dividing the length in its account, and, conceptually, becoming the limit of one and the beginning of the other, we make it many in account, though it is one in substrate; for it has one account taken as limit, another taken as beginning. But (2), in so far as we take it as
749,1 one not only in substrate but also in account, as holding the line together, it becomes in that way altogether one, i.e. not only in substrate, as in the case also when it divides, but also in account.

[225] Book 6.1
[226] Diels has only a comma here. (J.O.U.)

Alexander says: '"for thought" can be taken with "other" and not as applying to division. For for thought the dividing point becomes other, though clearly not in substrate.' But as the point is to the line in mathematics, so in natural science is the now to time. For as dividing time it is taken conceptually and, as it were, potentially. For it is not so in actuality and as remaining, but in this way the dividing point in the line, thought of twice, is potentially two, but becomes actually two when the parts of the line are separated from each other. But the now is potential as in that case, when each is taken twice in thought; for it never becomes so in actuality. For it does not remain. Alexander says: 'Perhaps potentially because through some potentiality and account the one in substrate becomes two in account.'

Such is the dividing now, but as limit, that is, as boundary and unification of both, it is taken as being in both alike. For when it has the same account in regard to both, then it unifies them and makes them one. But both the division and the continuity in both length and time are with respect to the same point and the same now. But the account of that which unites and that which divides is not the same, for the former is one and the same not only in substrate but also in account, the latter is taken as two in account.

But why did he use the division and continuity of mathematical magnitudes as an illustration, and not that of physical magnitudes? For in their case also what is said is true. Perhaps he used the mathematical illustration first because the now in the strict sense has no parts. But a point that has no parts is the mathematical one, not the physical; also, they can be divided only in thought, as he says is the case with time. Such a division is possible in practice in the case of physical things. Perhaps it is also possible in the case of time, if one has in mind the physical divisions of the parts of time, such as day and night, months and years, as is the case with change also. For I think there can be some physical divisions of this also, some in relation to the substrate, some through an actual difference within the change.

> **222a20-4** Nows are spoken of in this way, [but in another when the time of something is near. He will come now because he will come today; he has come now because he came today. But events in the *Iliad* did not happen now, nor is the flood now. Though time is continuous to them, they are not near....][227]

Having discussed the now proper, which is indivisible and without parts, and holds together and divides time, he adds also the other

[227] The lemma is stated to continue *to aei gar en arkhêi* at 222b7. In fact, only the passage to 222a24 quoted above is discussed now and a new lemma correctly begins at 222a24 on page 750, line 8. (J.O.U.)

750,1 senses of 'now', which we are accustomed to use, but loosely. For time close to a certain now, both past and future, being close to the now proper, is said itself to be now. For we say that the man who will come today will come now, and that the man who has recently come has come now. The events in the Iliad did not happen now nor is the flood now, although time is continuous. Things done then were not close to the now proper, but long ago.

222a24-b7 Sometime is determined [in relation to the earlier[228] now and the later now. Thus at some time Troy was captured and at some time there will be a flood. They have to be reckoned from now as a boundary. There will be so much time from this to that, and there was to what is past. If there were no time that was sometime, every time would be limited. So will it give out? Or will it not, if there will always be change? Is time then always new or does it often recur? It is clear that time will be as change is. If sometime one and the same change occurs, sometime will be one and the same, otherwise it will not. But since the now is the beginning and the end of a time, but not of the same, being the end of the past and the beginning of the future, time must always be at a beginning and at an end like a circle, which has somehow its convexity and concavity at the same place. That is why it always seems different; for a now is not the beginning and end of the same time. Otherwise opposites would hold of the same thing simultaneously. Also time will not give out;] for it is always at a beginning.

He will next try to teach about the adverb 'sometime' and other matters related to time. At the same time he articulates the things that have their position in relation to the now, aiming to show also by this means that no time occurs that has no limit. So he says: 'Sometime is determined in relation to the earlier now and the later now', as Troy was at some time captured. We say this, showing that there is a gap from the present now to the past now, which was when Troy was taken. It is the same in regard to the future; for we say

[228] The received text reads *to proteron nun, hoion* at 222a25. The Oxford translation and Ross in a note to his edition both say that *proteron* means the former kind of now, i.e. the now in the strict sense. But at 750,13-14 Simplicius has *estin oun, phêsi, to pote khronos hôrismenos pros te to proteron nun kai to husteron nun*. From this it is clear that Simplicius understood *to proteron nun* to be the earlier now, as I have translated it, and quite possible that his text contained also *kai to husteron nun*. To say that the date of Troy's capture has to be determined in relation to now in the strict sense seems ridiculous. Simplicius takes Aristotle to say that sometime is either measured from the present now to an earlier now (the fall of Troy), or from the present now to a later now (the flood). (J.O.U.)

that some time there will be a flood. From this he will finally show that any time taken is bounded.

Every time admits the predicate 'sometime'; for the point of 'if there were no time that was sometime' is not that a time was or will be, but that there is no time that does not admit the predicate 'sometime'. If every time is sometime, but every sometime is bounded (this was shown previously), every time is bounded. If any time taken is bounded, time will not cease, but will last for ever, since change, of which time is the number, will. It will be shown towards the end of this work[229] that change will never give out.

But if time is also the measure of rest, it will seem that time exists when there is no change. Alexander solves this difficulty saying: 'Time measures rest also, but its being and essence are in change.' For it is the measure of this, as was proved earlier. So let us recall that time should not be thought of as the number of change as distinguished from rest, but viewed as of that change which includes rest and all things that have from time their being in becoming throughout their existence. Its being and essence might be said to be in change because it measures the first and most important of all 751,1
changes, circular motion, and all other changes from that.

Having asked concerning time whether, as inexhaustible, it will be the same or different, he judged now also that, as in change, the same will recur over and over, but numerically ever different while the same in kind. For that is how it is with change. He shows clearly by means of the now that time is always ever different and is inexhaustible. For since the same present now, considered as present and not simply a limit, is simultaneously beginning and end of a time, the beginning of the future and the end of the past, and these are in the same now like concavity and convexity in the circle, it is clear that it is the limit of the one and the beginning of the other; for that is how it divides time. Also it is impossible for the same now to be the limit and the beginning of the same thing. Otherwise contradictories, which he called, more generally, opposites, would be the same thing at the same time. The beginning and the limit would be in a way opposites, and not merely as relatives are opposed, and a time contains opposites. If that of which the now is the limit is the past, and that of which it is the beginning the future, if the same now is taken as limit and beginning, past and future would be the same time, which is an opposition within a time. But if it is not of the same time, but the limit of the past and the beginning of the future, the future will necessarily be other than the past. For that very reason time will never give out, since the present now is a beginning just as it is a limit, the limit of past time, the

[229] Aristotle *Physics* 8.1.

beginning of the future. Therefore there will always be a future. For if it is a beginning, there must exist that of which it is the beginning, and inexhaustibly. That nothing prevents the same thing being beginning and limit, and in general opposites, taken in different respects, he showed in the case of the circle, where the same thing is at once all concave and convex, taken in different respects, and there cannot be the one line about the circle and not the other. Thus, once again, the now in the strict sense is both limit and beginning at once, through its relation to ever different times. For the present time, as such, is the mid-point between past and future.

> **222b7-12** What is already is that which is part of future time near to the present indivisible now [('When are you going?' – 'Already', since the time at which one intends to go is near), and the part of the past which is not far from the now ('When are you going?' – 'I have already gone'). But we do not say that Ilium has already been taken], because that is too far from the now.

'Already' is also a temporal adverb, as is 'sometime'. This also is 752,1 defined with reference to the indivisible now, not that in the broad sense. For we call 'already' what is in either direction near to the present now. For we say that which will soon happen or has just happened has happened already or will happen already. When asked 'When are you going?', we say 'Already' and 'I have already gone' when the time is near to the now. But nobody says 'already' of things that happened long ago. Thus nobody says that Ilium has already been taken, since it was taken long ago. The already[230] must be near to present time.

> **222b12-14** What is 'just' is the part of the past near to the present [now ('When did you come?' – 'I just came'], if the time was near to the present now).

He says that the 'just' is also a temporal portion of past time that is near to the present now. For if asked 'When did you come?', we say 'I have just come' if the time at which we came is near to the now. Aristotle understands 'just' as referring only to the past, as the example shows; but Alexander and Aspasius say that it can be used also of the future. For when asked 'When will he come?' we say 'He is just arriving' if he is about to arrive. 'Thus',[231] says Alexander,

[230] I have translated *êdê* as 'already', and am about to translate *arti* as 'just'. Simplicius' comments are more appropriate to the original than to the translation, as is also Aristotle's text. (J.O.U.)

[231] These opening quotation marks are missing in Diels. (J.O.U.)

'there is only a verbal difference between "just" and "already".' However, according to Aristotle 'already' is used in both directions, but 'just' of the past, and they seem to differ in that way. 'But neither of "already" and "just" is different from "now" as used broadly of a time near to the now in a strict sense', says Alexander. 'So', he says, 'all three of these expressions, "now" in the broad sense, "already" and "just", refer to the same time.' But Aspasius says that 'now' indicates being very close. Perhaps 'now' in the broad sense got so called because of its closeness to the now proper; but 'already', in addition to being near, has some terminal force, even if used of future time; but 'just' seems to indicate the newly, and might be used of the future also, but it is habitually used of the past.

222b14 The distant is long ago.

The part of time in the past that is far from the now was called long ago as the contrary of 'just', including in each case so complex a concept in one word.

222b14-15 The sudden is what is in a time imperceptible through its smallness. 753,1

'Suddenly' has two requirements: it must all happen at once in a short time, so that it seems to be in no time, which is what we call 'suddenly'; and, through the shortness of the time and the sharpness of the change, we must be unaware of its temporal duration. For that is why we say that lightning happens suddenly, because the burst of light occurs all at once, in a little time and sharply without our anticipating or getting used to it. That is why things that happen suddenly shock even the most fearless.

Alexander enquires how he can say that a time can be imperceptible through its shortness, while in *Concerning Perception and the Perceivable* he claims that of its own nature no time is imperceptible.[232] 'So perhaps', he says, 'since perceived and unperceived time are judged by a perceived or unperceived experience, when we are unable to apprehend some happening through the rapidity of its change and the shortness of the time in which it came about or changed, then the time taken by this event is unperceived, not totally imperceptible; for at the same time it is possible to perceive something else, if not this. For we are aware at least of ourselves at that time as existing and perceiving. For it is not possible to find a time at which we shall be unable to be aware of our own existence. So, absolutely, this short time must be

[232] Loc. cit. 446a29. (J.O.U.)

perceptible, but imperceptible in regard to this particular thing, which we say happened suddenly, so that the time in which such a change happened is called imperceptible because the change is unperceived. More colloquially', he says, 'he called the now that is the limit of a time, and is imperceptible in its own nature a time that has "just" occurred. For he is accustomed to call that also a time. Things that have happened at this time, which is the limit of changes, and, if anything, is instantaneous, happen at this imperceptible now, and these things are called sudden. For things that happen suddenly are called sudden.'

In certain manuscripts is written 'varied through its smallness',[233] which means 'changed'.

> **222b16-27** All change naturally causes variation. [But everything comes to be and passes away in time; that is why some have called it wisest of all, but Paron the Pythagorean most incapable of learning, since people forget in time, speaking more correctly. It is clear that in itself it is more the cause of passing away than of coming to be, as was also said earlier, since change naturally causes variation; but it is contingently the cause of coming to be. A sufficient sign of this is that nothing comes to be without it itself changing and acting, but a thing may cease to be without actively changing. This especially we usually call passing away through time. Not that time in fact brings this about,] but this change happens to occur in time contingently.

754,1 Having analysed the adverbs of time, 'now', 'sometime', 'already' and 'suddenly', he analyses further his earlier statement that time is more the cause of passing away than of coming to be, and also adds that it is not as such the cause of passing away. Thus, having premised that all change brings variation,[234] he makes use of this later in the demonstration. Meanwhile he says that since all that comes to be and passes away does so in time, some say, understandably, that time is very wise, others that it learns nothing.[235] Simonides says that it is very wise because we acquire knowledge through time, Paron the Pythagorean that it learns nothing, because people forget through time. It seems to be him to

[233] *ekstan* (varied) is supplied by Diels, necessarily. It is in some MSS of the *Physics*, and read by Ross in his edition. (J.O.U.)

[234] I have throughout translated *ekstasis* and cognates as 'variation' and cognates, not as ideal, but because 'change', 'alteration' and other more obvious words are needed for *metabolê*, *alloiôsis* and other terms. (J.O.U.)

[235] I translate the ambiguous *amathestaton* thus since the contrast is with forgetting things, not with knowing things. It might be better to translate *sophôtaton* as 'very knowledgeable' than as 'very wise'. (J.O.U.)

whom Eudemus refers without naming him; he says that at Olympia Simonides praised time as very wise, since in it learning occurs and recollection; but a wise man who was present said: 'What's this, Simonides, is it not time in which we forget?' But perhaps in Aristotle, in 'The Pythagorean Paron', 'paron' is not a proper name but the participle *parôn* [being present]; the Pythagorean being present when Simonides said that time was very wise said that it learns nothing. The latter seems to be more correct. 'All change naturally cause variation.' So it is clear that time as such is the cause of passing away rather than of coming to be, as was also said earlier. For both motion and change as such cause variation in a previous state, but since some things that are varied change from their own mode of being into others, through this change seems also to be contingently the cause of coming to be.

So, if time is the number of motion and change, while change and motion as such are the cause of passing away, then time also is as such the cause of passing away, since time is the number of change. A ground for the truth of what has been said is provided by the fact that for things coming to be time alone is not sufficient for them to come to be, but some activity is needed as well to bring them about, such as a skill, or nature, or teaching, or action, while for things passing away time on its own is sufficient for their demise, even if no external cause supervenes, as when things wear away or decay in time. We are accustomed to say that such things suffer thus through time. This is the passing away that happens through time.

Having said this, he adds an explanation why he has reverted to the earlier discussion. It is because time as such does not bring about even passing away, but this change also happens to occur in time. So what is the cause of passing away? 'The nature of each thing and the underlying matter', say the commentators. For there are measures determined by nature for everything in becoming for growth, prime and decline. That nature is the cause of growth and prime is a truism. But what is the cause of passing away? For we do say that things have a perishable nature. But nature in the prime sense of the word is creative and a cause of change in bodies is the principle of coming to be and pre-eminently the cause of being. That is why he says in Book 2 of this work[236] that the poet was ridiculous who said

> it has its end for which it came to be.

For nature does not look to passing away as its end, but to coming to be and existence.

Since perishable things come thus to be, one passing away as

[236] Aristotle *Physics* 194a31. (J.O.U.)

another comes to be, ceasing to be and the changes it involves will accompany artificial changes also. But we say that what is naturally perishable we declare to be material and imperfect. The cause of passing away, if one enquires, is the nature of sublunary matter,[237] which cannot permanently maintain its forms, and the disruption HERE through their descent of opposites that cannot coexist, as they can in their first intellective cause. If this is true, time should be pre-eminently the cause of coming to be and secondarily of passing away. But Aristotle here also makes use of the causes that appear most obvious, and holds time to be rather the cause of passing away. But he himself added 'rather', since he himself counted matter to be the cause of all change, because of its acceptance of opposites, as we learnt towards the end of the discussion of the void.[238] But how can time, since it is a measure, be said to be the cause of coming to be and passing away? Perhaps it is because the measure is a completion,[239] and, in the way that being three or four yards long is of a magnitude, so being for such or such a time is the completion of a coming to be or of a passing away.

> **222b27-9** So ends the statement that time exists and what it is, and of the different senses of 'now', and what sometime, just, already, of old and suddenly are.

It seems that the existence of time has not been demonstrated, but, on the contrary, it has been forcibly argued at the beginning of the discussion of time that it does not exist. But, in the course of showing what time is, by discussing an existent of which we have a concept, he at the same time proved that it does. He will also demonstrate it in what immediately follows, when he shows that every change and everything changing changes in time and that the before and after are in time.

CHAPTER 14

756,1 **222b30-223a15** Now that we have made these distinctions [it is clear that every change and everything changing is in time. For being quicker or slower is possible in every change, as is clear in every case. I say that that changes quicker which changes before another into a condition through the same interval and changing regularly. For example, in the case of

[237] Earth, air, fire and water, elemental matter below the moon.
[238] Aristotle *Physics* 217a21ff. (J.O.U.)
[239] *sumplêrôtikos*: completing the essence of something.

> motion, if both things move around the circumference or both in a straight line, and similarly in other cases. But now, what is before is in time; for we say that things are before and after with reference to their distance from the now, while the now is the boundary of the past and the future. So, since the now is in time, the before and after will also be in time. For the distance from the now will be in the same thing that the now is in. The before in past time and in the future are regarded in opposite ways; for in the past we call that before which is further from the now, and later that which is nearer, while in the future the nearer is before and the further is later. So, since what is before is in time, and every change involves a before,] it is clear that every change and every motion is in time.

He said a little earlier that everything both comes to be and passes away in time. He now proposes to prove this, offering a more general demonstration that 'every change and every motion is in time'. This proof closely concerns the enquiry into time; it also confirms what has been proved earlier, that time is the cause of passing away rather than of coming to be. The argument is as follows, set out in the first figure:[240] every change contains a before and after; what contains a before and after is in time; from which it follows that every change is in time. But, since the premises are not obvious, he provided a basis for each; for the minor premise: every change may be quicker or slower, for in every one a thing may change more quickly or more slowly; but everything that admits being quicker or slower admits being before or after, since by these the quicker and slower are determined. For we call that quicker which changes before into that into which the change is, and that slower which does so after. This distinction will be accurate if we take the same kind of change, such as change of place, and the interval equal and similar, such as round a perimeter or in a straight line, and if the change is regular. For if one changes regularly and the other irregularly, it is possible for what moves quicker to arrive at the limit of the change later. There may be a difference in the motion beyond the form of the substrate, if one should be straight, the other circular.

He demonstrated the major premise which says that the before and after are in time through the before and after being so called according to their distance from the now. For that is before which is further away from the present now in the past, while the after lies nearer to the now. But in the future, conversely, the before is what is nearer to the now and the after what is further. So the before and the after have their being in relation to their distance from the now. For the now is the boundary of the past and the future; 'So, since the

[240] The first figure is of the form MP, SM, therefore SP. (J.O.U.)

now is in time, the before and after will also be in time.' For the distance from the now, which is the before and after, will be in the same thing as the now is in. So, if every change contains a before and an after, and these are in time, every change will be in time, both premisses being taken as true. For he made it clear that he set out as closely related to the proof the premises, one saying accurately that every change contains a before and an after, and
757,1 that which says that the before and after are in time, by adding after the proof: 'So since what is before is in time and every change involves a before, it is clear that every change and every motion is in time.' 'Involves' means that the one follows from and is predicated of the other.

> **223a16-21** The relation of time to the soul is also worthy of examination, [and also why time seems to be everywhere, on land, in the sea and in the heavens. Is it because it is some affection or condition of change, being its number, and all those things are changeable, since they are all in place, while time and change] are at one, both potentially and actually?

Here he proposed to examine two questions, first, how time is related to the soul, i.e. whether there could be time if there were no soul, which is the same as to ask whether what is counted could exist without that which counts, since the soul is that which counts. But, though he set this question first, he will try to resolve it later. The other question is why time seems to be everywhere, on land, in the sea and in the heavens.

He explains this first, saying that time, as the number of change, is an affection or condition of it, so that it is an inseparable characteristic of change. For number seems both in things numbered and to be a condition of them. So, since all those things which are said to be in time, such as earth, sea and the heavens, are bodies, and every body is in place, and what is in place is changeable either as a whole or in its parts, and what is changeable possesses, together with change, its characteristic as affection or condition, which is time, therefore it stands to reason that in those things that admit change, land, sea and heavens, in them there is time also. Putting this succinctly, he added that 'time and change are together, both potentially and actually'. Having said that time is co-present with change, wherever change is present, he adds that time, in every case, is as change is; if change is potential so is time also; if it is actual, so is time actual. That is why, in drawing to a conclusion, he said that things in place were changeable, not that they were changing, so that both the potential and the actual were included. For both the changing thing and what naturally changes, such as what is at rest, are changeable, as Alexander says. He calls time, as a characteristic of change, an affection, since number is an affection of the numerable, and a condition because it is possessed[241] by change. But if what is at rest is now to be said to be potentially changing, as Alexander understood, while what is potentially changing is also potentially temporal, it is clear that what is at rest must be potentially temporal. But not only changing things but also things at rest were said to be actually in time. For time is a measure in actuality of rest also, since the duration of rest as of change is measured by time. So perhaps one should not understand things at rest to be potentially changing in relation to time, but things which have not yet come to be, but are capable of coming to be. For if time were the number of change in things that have come to be, and things said to be at rest were in that way changing, it is clear that change would be present actually in these things, and time would measure such change. But since Alexander understood rest to be in time contingently, he would say that change which is essentially in time was sometimes taken as potential, as in the thing at rest, but actual when the thing was actually changing.

Alexander reports another reading, perhaps a revision because of the above problem: 'place and change are co-present both potentially and actually', as a confirmation that the things mentioned, land, sea and the heavens, are changeable, by adding what is now said, that things that are actually in place are actually and essentially changing, while things potentially in place, as are the parts of a whole, are also potentially changeable. So the fact that, as each thing participates in place, it thus participates in change, potentially in each or actually in each, is an indication that things in place are subject to change.

The problem obviously arises how time is present in the heavens, since they are not in place,[242] given that only things in place are changeable and only changeable things are in time.[243] Alexander saw this and solved it, saying: 'It is very clear that time is present in the heavens, even if they are not in place; for they are in everlasting motion. But where there is change, there there is time. But, in the case of other things, that are not in everlasting motion, he needed place, since he showed that they were also changeable through their being in place. But the divine body[244] needed no proof that it was

[241] The translation fails to bring out the verbal affinity of *hexis* (condition) and *ekhetai* (possessed). 'Possession' as a translation of *hexis* would be too narrow for this technical term. (J.O.U.)

[242] Aristotle's doctrine is that place is the container of what is in place, but nothing is outside the heavens to contain them. (J.O.U.)

[243] Since they only revolve, the heavens do not change their place, according to this argument.

[244] The heavens.

moveable, for its motion, being everlasting, is clear and obvious.' These are Alexander's own words. But if the heavens move with a circular motion, it is clear that their change is of place and not some other form of change; but what changes its place is also in place.

> **223a21-9** One may raise the problem whether there would or would not be time if there were no soul. [For, if it is impossible for what will enumerate to exist, it is impossible for there to be anything enumerable, so, clearly, for there to be number. For number is either that which is enumerated or enumerable. But if there is nothing of the kind that enumerates except soul, i.e. the intelligence of the soul, it is impossible for there to be time if there is no soul, but only that of which time is an attribute, if, indeed there can be change without soul. But the before and after are in change and] time is these in so far as they are enumerable.

He has switched to the other of his proposed topics, which was stated first, and there as follows: 'what is the relation of time to the soul?', but is here more clearly restated. The question is whether or not there could be time if there were no soul. For it is plausible that related things exist together and are annihilated together (if there is no right there is no left either); but the enumerator and the enumerable are relatives, and soul, i.e. the intelligence of the soul (since we do not enumerate with our feelings, but with reason), enumerates. But time is something enumerable, so that it seems that without soul there can be no time.

He opposed that which will enumerate to the enumerable; but, they say, one ought so to oppose the enumerator. But the enumerator and that which will enumerate are the same thing, the one being taken potentially, the other in relation to future time, which is equivalent to the potential, so that the enumerator and that which will enumerate are the same thing. He added that number is either the enumerable or the enumerated; for we call number either what has already been enumerated, or what is capable of having been enumerated, as the mass of soldiers is enumerable because it is capable of having been enumerated. What has been enumerated is also enumerable. So, if there is nothing enumerable there can also be no number. So the entire process of drawing to a conclusion in three stages runs as follows: if there is nothing which will enumerate nor is there the enumerable; but if there is nothing enumerable there is no number *qua* enumerable. So if what will enumerate is the soul through its intelligence and time is number, if there were no soul there would be no number.

Boethus[245] attacks this argument, saying that nothing prevents there being the enumerable quite apart from an enumerator, as there can be the perceptible without a perceiver. But Alexander also stated the attack in several stages, and added a solution in accordance with Aristotle. He says: 'It is worth investigating how sound the statement is that there is nothing enumerable if there is nothing that will enumerate. That there is nothing enumerable or enumerated *qua* enumerated if there is nothing that enumerates can be allowed, if being enumerated is essential to number. But what is enumerable itself, i.e. what is capable of being enumerated, such as men or horses, does not seem to be annihilated with the enumerator. Aristotle himself shows that the before and after in change, which are enumerable, can exist without soul, at least if there can be change without soul.' Solving the difficulty he says: 'Perhaps the enumerable will not exist, but that which is contingently capable of being enumerable will exist, such as horses or men, but not the enumerable *qua* enumerable. For in the case of other relatives, if, for instance, there is nobody on the right, he who was on the left, Socrates perhaps, will exist, but a person on the left will not. So, if time were enumerable as the before and after are enumerable, if there were nothing to enumerate there would be no time. But nothing prevents the substrate of time, which is change, from existing.' Aristotle provided this solution succinctly by 'but only that of which time is an attribute'. For nothing prevents the substrate of time, or the enumerated, from existing without soul existing, if it is in general possible for change to exist without soul. For if things actual are taken in relation to other things actual, and potential things to the potential, if it were impossible for that which will enumerate to exist, the enumerable would also be impossible taking the enumerable to be what can be enumerated. Again, if perception were abolished, while the colour would remain, it will not exist *qua* visible, but simply as colour.

He very properly added 'if it is possible for change to exist without soul'; for, so far as the antithesis of relations goes, if soul be abolished for the sake of argument, nothing prevents there being change, even if not time. If, however, it is not possible for there to be change without soul, not only time but change also will be abolished in the absence of soul. For if the cyclic motion, from which all other motions and changes derive their existence, is governed by intelligence and design, as Alexander also agrees, if soul were abolished all change would be abolished. For local motion depends on cyclic motion,[246]

[245] An Aristotelian of the first century BC whose work on time is identified by Pamela Huby, 'An excerpt from Boethus of Sidon's commentary on the *Categories*?', *Classical Quarterly* 31 (1981), 398-409.

[246] Of the heavens, as opposed to the rectilinear motion of the four simple bodies,

since the changes of simple bodies into each other derives from it, as they do through heaviness and lightness, and local motion is natural motion, unnatural motion subordinate to natural motion. The natural change of animals is caused by their own soul and cyclic motion. 'For man begets man together with the sun.' So the cyclic motion is the cause of that change which is birth, and the changes through alteration and through growth and diminution are from it also. For the approach and retreat of the sun provides the major explanation of these things.

When this has been said, it is worth noting, I think, that if the enumerator is abolished time as an enumerated number is abolished, but as the before and after in regard to the duration of change it is not abolished. In that respect time exists. For it is, with regard to the before and after, contingent that it is enumerable. But if soul is abolished, which is the principle of becoming and of all change in regard to becoming with which time is co-present, time also will be abolished. So Aristotle reasonably says that, if the soul *qua* enumerator is abolished time will be abolished *qua* enumerable; but he shows that, if it were abolished *qua* the principle of coming to
761,1 be and change, both change and coming to be would be abolished, by 'if it is possible for there to be change without soul'. But not only is time, being an aspect of change abolished together with the abolition of change, but, since it itself has its being essentially in becoming, it will be abolished together with the abolition of soul, which is the acknowledged cause of all change and becoming.

Note how here, also, Aristotle wrote in accord with our leader.[247] He himself wrote 'if it is possible for there to be change without soul', i.e. without the changer, while the other said that for other things that change the soul is the 'source and principle',[248] and that from this principle everything in becoming comes to be.

> **223a29-b1** One may raise the question also of what kind of change time is the number. [Is it of any kind whatsoever? For things come to be in time, and perish and grow and alter and move. It is the number of each change *qua* change.] So in general time is the number of continuous change, but not of any one.

He has shown that time is the number of change, and, up to now, stated most of the argument as applying to change of place. To this is suited the statement that change was measured by its magnitude and his saying that 'we say the road is long if the journey is long and

earth, air, fire and water.

[247] Plato.

[248] Plato *Phaedrus* 245C. (J.O.U.)

that this is long if the road is long'.[249] But there are many kinds of change, change of place, of quality and of quantity, and there is also the substantial change of things coming to be and passing away; though this is not change, as the book will later show for the precisian; but still there is a before and after in this also and participation in time is seen. He therefore rightly says it is worth questioning of which of these changes time is the number. He says that it is not of some one, as such, but of any sort of change. It is not because it is passing away or alteration or some other change that time is its number, but as being change and change as such, which is common to all the changes mentioned. So time is the number of all change equally, since a before and an after can be seen in all of them, and time is that which is numbered. So change is in time, not as being this sort but simply as change.

Having well said that time is the number of change, he added 'continuous, and not of any one'. For that which is numbered in time contains a before and after in so far as change is continuous. In the 762,1
case of discrete change, it is measured by number and not by time. For two, three or more changes take place measured by abstract number. Thus if one is alteration, another passing away and another growth, since none is continuous with the others, they are three changes. But in so far as each of them is continuous, their befores and afters are numbered, not this time by abstract number, but by time.

Alexander says that the before and after of all change provides the existence of time in so far as it is continuous and is consequent on the magnitude of that in which it occurs, in which there is first of all a before and an after. One may raise both for him and for Aristotle a difficulty which, I believe, I earlier raised in short form and Themistius at greater length. 'For if the before and after are first in the magnitude and the interval over which the change occurs, and thence in the change and the time, how are the before and after in growth and alteration? For they do not occur in an interval nor are from here to there, but the body as a whole warms up gradually, and grows and solidifies gradually. Whence, then, in these cases will time get its before and after, as it does in the case of motion from interval and position? How, not having a before and after, will it have a number in respect of them? Perhaps some one may say that in the case of growth there is a from here to there; for the change is from a smaller to a larger place. But what shall we say for the cases of alteration, coming to be and passing away? Whence now will the change or the time get its before and after? So either time measures pre-eminently only change of place, or, if the others also

[249] At *Physics* 220b29. (J.O.U.)

pre-eminently, it has no need of a before and after in position in order for it itself to have a before and after.'

But, since Themistius also seems to have conceded this difficulty, bidding us to examine it repeatedly and not to admit it without examination, perhaps there is no objection to my stating now my view, as well as the problem, and submitting it to the examination of such as encounter it. In my opinion, the before and after which exhibits order is one thing when viewed in relation to time as the substrate of the many nows that in number and order determine time, and another thing when viewed as what has a position in an order. The latter is twofold, first in magnitude (as that portion of the 763,1 road from which we start is before, and that in which we finish is after), second in number. This is also twofold, either as the abstract number that enumerates (as we say that one is before two, and two before three), or as that applied[250] and enumerated (as intelligible things are before the perceptible and this pillar is before that; for incorporeal things also seem to have a position in a metaphorical sense of 'position'). A third sense of 'before' and 'after' is in relation to activities and changes, first in regard to the number of the limits bounding a motion, which we call starts analogously to the points in a line and the nows in time, second, in relation to the parts of the change which is continuous in itself. But a change in place, which receives in itself the before and after in regard to position in place, harmonises this with its own before and after, both as a limit and as a part, as when we say that the first motion of the runners is where the officials are, the latter that which is at the end of the track. This before and after in position belongs to change of place. For this and for the other changes that have their being in becoming time provides from itself a before and after which it harmonises with the before and after of the change. For the nows that determine time which occur within the bounds of a change determine the change within time. So time is the measure of change. For it measures the change by harmonising its own before and after with the before and after of the change. Also that is how change becomes the measure of time, when the befores and afters belonging to change, harmonising with the befores and afters of time, determine the time by the change. In that way also spatial magnitude measures, not all change, but change of place, when the before and after of place in regard to position harmonises with those of the change. But if Aristotle treats the measurement of change by magnitude as universal, he probably does so because of its obviousness. That is why he continues for most of the time to argue as if it was about change of place.

[250] Literally, participated.

223b1-12 But it is possible that another change is occurring[251] now. [So there should be a number of each change. So is there, then, another time, so that there would be two equal times? Surely not: for the equal and simultaneous is one and the same time. Those not simultaneous are the same in kind. For if there were dogs and horses, seven of each, there would be the same number. In that way also there is one time of simultaneous events with the same limits, though perhaps one is fast, the other not, one a motion, the other an alteration. But the time is the same, since it is equal and simultaneous, for both the alteration and the motion. For that reason the changes are different and separate, but time is everywhere the same] because there is one number which is the same everywhere for equal and simultaneous things.

He sets out a problem consequent on what has been said about time, and at the same time regards his statement, often made, that there was the same time everywhere as requiring further articulation. The problem is as follows: if time is the number of change and change is in that which changes, as we have learnt, and if time is dependent on change, there should be as many times as there are changing things and changes. So there will be many times simultaneously, and not just as contained and containing; for there is nothing odd about that.

Solving the problem, he says that this is not the case. For there do not come to be many times simultaneously, but every equal time present is numerically one and the same, while those also that are not simultaneous, i.e. the past and future, are the same in kind. After this, he explains how, when there are many changes, there is one time, although it is the number of change. For just as, in the case of abstract number, the number is not divided up together with the difference in the things numbered; e.g., if of the things numbered some are horses, some dogs, seven of each, they differ from each other, but the number seven is the same in both cases (for they do not have the same number *qua* different, but through being so many, which is common to them): time is, in the same way, the number in the case of changes. For the now is numerically one and the same for all the changes, even if one is local motion, another alteration, while what is numbered and taken as before and after in all of them is the now. So, if this now is the same for all of them, time is, as one would expect, the same for all of them. If time occurred differently in changes, like being white and being hot, it would, reasonably, be divided together with them; but, since what is numbered in them is a quantity, nothing prevents this from being

251 Simplicius reads *kineisthai*. Ross accepts *kekinêsthai*. (J.O.U.)

the same for many changes, just as, in the case of abstract number, what is numbered is the same, being so many, even if many things differing from each other are numbered. Thus there is the same time for simultaneous changes, but not the same speed nor the same place; for the numbered changes do not create time by their quickness or slowness, nor by their difference in kind, as when one is alteration, another local motion, another growth, but by the befores and afters alone, which are the same for all changes which being simultaneous and so conceived are now. So, if the nows by which we number time are the same, the time, which is determined and bounded by these nows will be the same. For, just as a difference in things changing place, if it is in some other way, such as colour or taste, and not in respect of that motion, like lightness[252] or slowness, does not make a difference in the motions, so the differences in changes also do not make a difference in time, since time is their number, not in the way that they differ, but in accordance with their before and after, which is the same for all; for their number is the same.

'But number', says Alexander, 'is not an element in what is numbered; so that time, being a number, is also not thus in change. Neither is the name in that of which it is the name in the way that white may be, because the numerically identical name belongs to many things simultaneously, as is the case with number, but white is not identically the same in many things.' Alexander added this to show that time is not in change in the way that the number marking its divisions has its being in being numbered. For, as Aristotle showed, without an enumerator there will be no number that is time, even if there be change. But, if time gets its existence from the enumerator, and thus comes about in change, it will not belong to it as an element in it. Therefore it does not mark its divisions. On this matter one must ask Alexander how the number in what is numbered does not belong to the enumerable, or how time that is the number of change as the enumerable does not belong to change nor is an element in it, although it is said to be an affection and condition of change. For he said 'or is it because it is an affection or condition of change, being its number'.[253]

Still more, how is it reasonable that the enumerated number, and the condition or affection, or, in general, any attribute should be related to that of which it is so, as is the name, so that the numerically identical belongs to a plurality? For the enumerating number, being outside the things enumerated, can apply to many as one and the same; but the enumerated number, fixed in its

[252] Lightness is a natural tendency to move upwards, by definition. (J.O.U.)
[253] Aristotle *Physics* 223a18. (J.O.U.)

possessor and coordinated with it, has the same status as white. But this name, to which Alexander digressed, is conventional and an irrelevancy. It is absurd to think that things conventionally imposed are akin to what is correlated by nature. In this he followed those who say that number is brought into existence by him who enumerates and that time is in this way brought into existence by soul, which was conjectured because the enumerator and the enumerated are correlatives. As enumerated, let it be so, as is colour *qua* visible; but in its own nature it exists independently. For Aristotle himself earlier chided those who defined colour as the proper sensible of sight, but did not exhibit its own nature. I think that in the same way one may chide one who said that the applied number had its being simply in being enumerated. For the fingers of my hand are five, even if nobody enumerates them.

But Themistius well raises a problem about Aristotle's statements, and it will be sufficient to hear his comments: 'For if', he says, 'time is the common number of every change, it seems not to get its being from change; for, if not from this or that, then it is not from change in general. For it is absurd to say that the before and after is the same in all changes. So one must investigate this very 766,1 point, in what way the now is identical, whether in genus, in species or numerically; for, if it is so in genus or species, it must be many numerically; but, if it is one numerically, how can things numerically different have a numerically single affection or condition? For that is how he describes the relation of time to change. For it is impossible for what is numerically one and the same to belong thus to many. In that way there will be simultaneously one and not one, so that, if there are many nows for the many changes, there will simultaneously be many nows for changes that are simultaneous, which is inconceivable. But, if there are not many nows, but one and the same, it will be separate and time will be no element in change. But this also has no plausibility, since, as the enumerated number is the same in ten horses and ten sheep, so it is also in changes. Nor will the enumerated number be the same there, except verbally that "both are ten", which is our concept and view of what is similar in different things; but there will be nothing substantial. But how, in the case of time, is it plausible that time is merely a concept of our souls, and has no nature of its own, as Aristotle seems to admit when he agrees that there will be no time if there is no soul. For to say that time is a measure and a number is to give in to that belief, as Boethus also says, since nothing becomes a measure in nature but measurement and numbering are our handiwork.'

That is what Themistius writes. I think that he has set out the argument well in other respects, for how can time be an affection or condition of change and, further, if separate, be the same simultaneously everywhere? But to say that the universality of applied numbers is merely conceptual seems to be a Peripatetic view, and needs, I think, some qualifications. For this universality is also real, which is why the things to which number is applied have a common nature. But it is real together with difference; to see the universality in itself belongs to our conceptualisation. But neither does Aristotle deny that time has its own existence, nor does he make it conceptual, when he says that there would not be time without soul. For he thinks that change is the substrate of time and that soul contributed the number as a kind of form, since soul is the cause of all becoming and change, and so reasonably says that if soul be abolished so will time also. But he does not therefore imply that time is merely a concept, because it is abolished together with its cause.

How, also, is there no natural measure, since all natural kinds are
767,1 structured by their own measures and numbers? Unless one were to say that there is nothing that is merely a measure without there being something measured, since there is no natural yard or unit existing as a pure measure.[254] If this is so, one must understand that even if it be the measure of change, it is also itself measured by change.

It is worthwhile to examine where these things are well said, or not, by these men. As things are, how does Aristotle apparently say both things about time, opposite to each other; both that it is an affection or condition of change, as inseparable, and that it is one and the same in different changes, as separate? Perhaps he hypothesised that time was altogether number, and saw in it the property of number, present even in applied number, that, even if the substrates be different, the number is unchanged, as the unity of a horse and the unity of a man can be abstracted in thought from their substrates and are altogether identical. But the whiteness in snow and that in white lead, even if abstracted in thought from their substrates, differ from each other by a difference in whiteness. For there are differences in whiteness and sweetness and other attributes in themselves, even if they be taken without their substrates, but there are no differences in equal numbers. For unit does not differ from unit *qua* unit, not even when applied, nor does quintet differ from quintet, as such, just as in the case of abstract numbers one can find no difference between the units. Aristotle himself will say something about this later on. But Iamblichus

[254] LSJ translates *autometra* as 'self-measured', referring only to this passage. That is not what is needed. I take *autometra* as a noun, akin to Aristotle's *autoanthrôpos*. (J.O.U.)

thinks that he is here retreating from inseparable time to separate time, and that that is why he says that it is numerically the same in the different changes. One must also remember the earlier solution, that time is the number of change through the definitional unit of change and of its duration, which is the same everywhere.

> **223b12-21** But since there is motion, including circular motion, [and everything is numbered by some one thing of the same kind, units by a unit, horses by a horse, so is time by some defined time, and time, as we have said, is measured by change and change by time (i.e. the quantity of change and of time is measured by the time of some determinate change). If, then, what is first is the measure of all things of the same kind, uniform circular motion is the chief measure, because its number is most well-recognised. Alteration, growth and coming to be are none of them uniform], but motion is.

He has said that time is number *qua* enumerated and a measure *qua* measured. But since everything that is numbered and measured is numbered and measured by something of the same kind, as units by the unit, yards by the yard, horses by a horse, and, in general, by the simplest proper part of the whole class (for Plato also says that there are as many measures as parts[255]): for even if they be measured by some other thing, they are through it also measured by the part, because the measure is equal to a part of the whole – since, then, as has been said, what is measured is measured by some portion of itself, and time is measured as something continuous, it follows that time is measured by a portion of itself. He therefore looks for the primary and most simple time, which will measure all time. So, if time is the measure of change, if some change be found which is the measure of all other changes, it is clear that its time will be found to be the measure of time. So what change is such? Clearly, that which is primary, uniform and well-known. For one that was not primary and was irregular and unknown would be an unsuitable measure. But the proper changes of generated things, alteration, growth, decrease, coming to be and passing away and rectilinear motion, are not prior (for the generated is not prior to the eternal), nor are such changes uniform, but they exhibit every kind of irregularity. For alterations, increases and comings to be are not equal in an equal time. Nor are they well-known;[256] only circular motion has this character; so this is the

[255] Diels writes 'nescio ubi'. Nor do I know where. (J.O.U.)

[256] I have here, as elsewhere, translated *gnôrimos* as 'well-known'. But 'easily recognised' or even 'intelligible' might be better. Compare Aristotle's *gnôrima haplôs*,

measure of change and, consequently, of time. It becomes the measure of change by being determined by time and thus being a certain quantity. For example, spring is a part of time, and determines a quantity of change in the rotation of the universe. Similarly a day and a year, by determining the change in the motion of the universe and of the sun, provided a measure of all changes and of the whole universe. For the prior and lesser time measures time; the hour is the measure of the day, the day of the month and the month of the year. They measure the hourly and daily change. But if these are such a quantity of the revolution of the sun, it is clear that they, in turn, are measured by change. That is why Aristotle said that time is measured by change, and change by time. Through a bounded change both change and time are measured by time. So the time that is reciprocally measured by the cyclic motion, which is primary and uniform, measures the whole of time. So, if everywhere the primary measure is of things of the same kind, it is clear that in change also cyclic motion, being primary and uniform, is the main measure, both for those reasons and because its number is most well-known.

Alexander understood the primary[257] time to be the number of cyclic motion, as being the best-known time; for the number of cyclic motion is days and nights, and this is the best-known of times. 'Perhaps', he says, 'he said that the number of uniform change was the best-known because that which numbers this and that which has its being in being numbered, which is time, are equal. In this motion one cannot find one faster, one slower, whose times are equal. So the uniform motion that is determined by the best-known time is the measure.' But I think that it is possible to say that the number of cyclic motion is well-known, not meaning time, but with regard to how this cyclic motion is numbered, as that the universe has revolved twice or three times. For this is easily assimilated and more relevant to one who is engaged in showing that cyclic motion was the measure. The time which determines this motion is best-known and primary; for we measure a month by days, and a year by months and the whole of time by years. It will be shown in the last book of this work[258] that cyclic motion is the only uniform motion.

One might raise a problem which seems to have exercised Plotinus also. If time is the number or measure of uniform and continuous change, what would be the measure of irregular and disjunct change? But if time is the number of all change equally,

which, contrasted with *gnôrima hêmin*, are certainly not well-known, but intrinsically intelligible. (J.O.U.)

[257] Reading *ton prôton* for *tên prôtên* at 768,28-9. (J.O.U.)

[258] Aristotle *Physics* 8.7ff.

how again will it be taken as numbered and not as separate, as one has the same dozen of elephants or ants, since as numbering it would be separate? Perhaps, as uniform, continuous and ordered change has a pre-eminent existence, from which irregular and disorganised change is derivative, so uniform change is measured pre-eminently by time and irregular change in a secondary sense. For we measure this also by hours and days.

It is noteworthy that Alexander wrote 'this is because it is by the time of a determinate change', and says it was said in place of 'by a determinate change by time'. Some of the transcriptions have the reading 'by a determinate change' on the assumption that people may perhaps have made the addition for clarification.[259]

> **223b21-4a2** That is why time seems to be the motion of the sphere, [since other changes are measured by it and time is measured by this motion. This is the origin of the stock saying that human affairs move in a circle, as do other things that suffer natural change, coming to be and passing away. This is because all these matters are judged by time, and begin and end as if forming a circle. Time itself seems to be a kind of circle, and this in turn seems so because it is the measure of such a motion and is itself measured by such a motion. So to say that the passage of affairs forms a circle is to say that time itself is a sort of circle, and this, in turn, is because it is measured by cyclic motion. For nothing appears in what is measured except the measure,] the whole being a plurality of measures.

Aristotle was accustomed often to help those who were seeking a knowledge of the truth but not grasping it accurately by giving a reasonable explanation of their misunderstanding. At the same time he puts his own account together by showing that those who made mistakes did so while aiming at something similar. He first compares those who say that time is cyclic motion: this has a certain point, since by this all changes and time are measured. Then he shows that 'the stock saying that human affairs move in a circle', and even the affairs of everything that comes to be and passes away, accords with what is said about time and circular motion, because all of these are judged by time, and have their own temporal beginning and end. That time is a circle is because it is both itself a measure of circular motion and is measured by circular motion. So to say that affairs are also circular is the same as to say that there is

[259] Alexander read the first: *touto de estin hoti tês hôrismenês kinêseôs khronôi*. Simplicius the second: *hupo tês hôrismenês kinêseôs hupo khronou*. Ross the third: *hupo tês hôrismenês kinêseôs*.

a certain circle of time, and this because it is measured by circular
770,1 motion. So, if what is measured seems to differ in no way from what measures, except perhaps in quantity (for that which is measured consists of many measures, and what is measured by the yard differs from the yard by being many yards long), then time also, being measured by a circle, should be a circle, many times repeated. For the substrate of measures is different in what measures and what is measured, as when the measuring yard is made of bronze but what is measured is made of wood; but that by which one measures, the other is measured, is the same yard, i.e. such and such a length, or, similarly, a circle of such and such a size.

> **224a2-17** It is rightly said [also that the number of sheep and of dogs is the same if the two are equal, but it is not the same group of ten or the same ten, just as the equilateral and the scalene triangle are not the same triangles, although the figure is the same, since both are triangles. For things are of the same kind if they do not differ by its differentia, but not of the kind in which they differ. Thus one triangle differs from another by the differentia of triangles; so they are different triangles. But they do not differ as figures, but are in one and the same division. For one figure is a triangle, another a circle, and of the former one sort is equilateral, the other scalene. So the figure is the same, being triangular, but they are not the same triangle. And so the number is the same, since their number does not differ by the differentia of number, but they are not the same group of ten. For the group referred to is different, in the one case dogs, in the other horses.
>
> That concludes the investigation of time itself and of matters germane to it.]

What is said here is, like what was said earlier, that just as the number of equal and different numbers is the same, e.g. the septet of seven horses and seven dogs, so time also is a number that is one and the same for different simultaneous changes. Aristotle laid this down again, not because he was given to excess verbiage, but because he is trying to articulate more exactly about it, as being open to many objections, and at the same time he has set it out together with parallels in other areas.

It seems quite clear that there is the same number of the seven dogs and of the seven horses,[260] even if they differ in substrate. I judge that it is from this consideration that he resumed the subject,

[260] It is clear that either Simplicius, or his editor, Diels, has omitted *kai tôn hepta hippôn* – 'and of the seven horses' –after *kunôn* – 'dogs'. (J.O.U.)

and from his saying that 'it is rightly said that the number of sheep and of dogs is the same if the two are equal, but it is not the same group of ten nor the same ten'. 'He is teaching us', says Alexander, 'by this how we should judge which things falling under something identical are different from each other, and which are the same. For the things that do not differ by the differentia of that by which they are called this or that are, in that respect, the same as each other; but things that differ by that thing's own differentia are, in that respect, different. For the equilateral triangle and the scalene are both of them figures, and both are triangles. But they are the same figure as each other, since they do not differ by the difference relevant to figures; for these differentiae are being rectilinear and bounded. Triangles all satisfy the same differentia of a figure; for they are rectilinear and moreover enclosed by three straight lines. But they are not also the same triangle, for they differ from each other by the differentia of the triangle. For of the triangle as such the equilateral, the isosceles and the scalene are different kinds. Also the number of things equal in number and not differing by the differentia of number is the same in respect of number, like the septet of sheep and of dogs. For the differentia of number is being odd and even, and each of these consisting of so many units. The seven applying to the things that have the same undifferentiated number is the same in essence. But the things numbered, such as the seven sheep and the seven dogs and the groups of seven they form, are not the same. For the sheep and the dogs, the things numbered, are not the same. Thus also a man and a horse are the same kind of substance, for they do not differ from each other by the differentia of substance, since the different substances are the living and the lifeless, and the things mentioned are substance *qua* living. But they are not the same animals, since they differ from each other by the differentia of animals. For being rational or arational is a differentia of animals. Again, Plato and Socrates are the same, *qua* animals, since they do not differ from each other by the differentia of animals, but they are not the same men, for these two differ from each other in the way that men are naturally different.

'It is the same regarding tens and tens; the number is the same, since tens do not differ from tens by the differentia of number. They are both even, and they do not even differ by the differentia of even numbers, which must be being composed of more or less units. But the substrates of which they are predicated, and these are the things that are numbered, are different in genus, or species, or number, and therefore no longer the same groups of ten nor the same tens. For the things numbered can be spoken of both ways.

'The present time is also the same everywhere; for it does not differ by the differentiae of time, being simultaneous in Europe and

Asia. For the differences in time are only being past and future. The different times do not depend on differences in changes. We do not number simultaneous changes as being at different times. But the changes of which time is the number are not the same, if one is an alteration, another motion. For these are different kinds of change.'

On this topic one must understand, first, the way in which this rule of identity and difference differs from that usually stated to distinguish identity and difference in genus, species and number as being one or many. In this latter case we say that things of which the same genus and species is predicated are one and the same in genus and species, as a man and a horse are one in genus, since both are animals, while Socrates and Plato are one in species, since both are men. That of which the same individual is predicated is one and the same in number. But by the rule now put forward, if the same word is predicated of two or more things, they are said to be the same in that respect when they do not differ by the differentia of that kind of thing, different when they so differ. For the equilateral and the scalene triangle are the same figure, since they do not differ by the differentiae by which figures are divided, which are being rectilinear and bounded and, if you wish, the conjunction of these; for both the triangles are rectilinear and consist of three straight lines. But they are not the same triangles, but different, since they differ by the differentiae of triangles, which are being equilateral, isosceles and scalene. For triangles are divided by these differences. 772,1 So, where by the first rule things would be said to be the same in genus and species, here some difference is found, because there identity was affirmed on the basis of sharing a common genus and species, and difference on that of being in different genera and species, but here identity is confined to sharing the same differentia of a common predicate; for we say that all equilateral triangles are the same and all rational animals the same. But there is difference in some respect when things differ in respect of the differentiae of the common predicate. For the equilateral differs from the scalene as triangle from triangle, since they differ from each other by the differentiae of triangles. So there identity and difference depended on genus and species, here on differentiae. Perhaps it depends on species immediately; for satisfying different differentiae of the same genus makes the species different.

From what has been said I think that it has become clear how Aristotle says that it is the same now at the present time, even if different events occur at it, since it is not different by the differentiae of the now, as the earlier and the later are different by the differentiae of the now and are so not the same now. This is the way that the equilateral and the scalene are not the same triangle, since they differ by the differentiae of triangles, but equilaterals are the same, since they do not differ by the differentiae of triangles. However, these too are different in number, not by differing by the differentiae of number, but as being separate individuals. But if the individuals differ from each other by no differentia, since they are nothing but individuals, like the individual units of abstract number, they cannot be said to differ from each other even in number. For Socrates and Plato are said to be the same in genus and species, since both are animals and men, but they are said to be numerically different, since they are different individuals differing in many attributes, since one was the son of Sophroniscus and Phainarete, the other of Ariston and Periktione. And one was snub-nosed, with protruding stomach and bulging eyes, the other had a shapely nose, was broad of body, and had well-set eyes, and one was the teacher of Plato, the other the pupil of Socrates. Their activities were also different. So the individuals now, taken as being in the present, are one and the same without difference, but in abstract number they are different, and if someone seeing them with their substrates were to take the differences of their attributes into account, he would say that they differ numerically. The earlier and the later now, however, differ in respect of the differentiae of their species; so that when Aristotle says that nows are the same, or time, when he says[261] 'it is the same time everywhere', and, elsewhere, 'the same time is that which is equal and all simultaneous, but time not simultaneous is one in species', it is clear that he calls one and the same that which does not differ by the differentiae of the now. So with time that is always equally present. But earlier and later are one only in species, not because they differ by the differentiae of the same species, in which case they would not be one and the same, but because they are of the same species either as now or as time.[262]

[261] Aristotle *Physics* 220b5, 223b3. (J.O.U.)

[262] The next sentence reads 'that must suffice to clarify what Aristotle says about time'. It is treated by Diels, the editor of the Greek text, as being the first sentence of Simplicius' independent discussion of time, usually called 'Corollary on Time', which follows immediately. (J.O.U.)

Bibliography

Place

Myles Burnyeat, 'The sceptic in his place and time', in Richard Rorty, J.B. Schneewind, Quentin Skinner, eds, *Philosophy in History*, Cambridge 1984.

E.R. Dodds, *Proclus, the Elements of Theology*, Oxford 1963.

Pierre Duhem, *Le Système du Monde, Histoire des doctrines cosmologiques de Platon à Copernic*, 10 vols, Paris 1913-1959 (see esp. vol. 1, chs 4 and 5 for Antiquity).

David Furley and Christian Wildberg, tr., *Philoponus: Corollaries on Place and Void, with Simplicius: Against Philoponus on the Eternity of the World*, London and Ithaca N.Y. 1991.

Edward Grant, *Much Ado About Nothing, Theories of Space and Vacuum from the Middle Ages to the Scientific Revolution*, Cambridge 1981.

Philippe Hoffmann, 'Simplicius: Corollarium de loco', in *L'Astronomie dans l'Antiquité Classique, Actes du colloque tenu à l'Université de Toulouse-le-Mirail, 1977*, Paris 1979, 143-63.

Edward Hussey, *Aristotle's Physics, Books III and IV*, Oxford 1983.

James G. Lennox, 'Theophrastus on the limits of teleology', in W. Fortenbaugh, ed., *Theophrastus of Eresus, On His Life and Work*, Rutgers University Studies in Classical Humanities 2, New Brunswick and Oxford 1985.

G.E.R. Lloyd, 'Right and left in Greek philosophy', *Journal of Hellenic Studies* 82, 1962, 56-66.

Concetta Luna, 'La relation chez Simplicius', in I. Hadot, ed., *Simplicius – sa vie, son oeuvre, sa survie*, Peripatoi vol. 15, Berlin 1987.

Henry Mendell, 'Topoi on topos: the development of Aristotle's concept of place', *Phronesis* 32, 1987, 206-31.

G.E.L. Owen, *Logic, Science and Dialectic: collected papers in Greek philosophy*, London 1986, v. index s.v. 'Location'.

Marlein van Raalte, 'The idea of the cosmos as an organic whole in Theophrastus' *Metaphysics*', in W. Fortenbaugh and R.W. Sharples, eds, *Theophrastus as a Natural Scientist*, Rutgers Studies in Classical Humanities 3, New Brunswick and London 1987 with contrasting assessments by Glenn Most and John Ellis.

S. Sambursky, *The Physical World of Late Antiquity*, London 1962, ch. 1.

S. Sambursky, *The Concept of Place in Late Neoplatonism*, Jerusalem 1982 (introduction and texts with translation).

L. Schrenk, 'Proclus on Space as Light', *Ancient Philosophy* 9, 1989, 87-94.

——, 'Proclus on Corporeal Space', *Archiv für Geschichte der Philosophie* (forthcoming).

L. Siorvanes, 'Proclus on the Elements and the celestial bodies: physical thought in late Neoplatonism', PhD diss. London 1987.

Friedrich Solmsen, *Aristotle's System of the Physical World*, Ithaca N.Y. 1960, ch. 12.

Richard Sorabji, *Matter, Space and Motion: Theories in Antiquity and Their Sequel*, London and Ithaca N.Y. 1988.

J.O. Urmson, tr., *Simplicius: Corollaries on Place and Time*, London and Ithaca N.Y. 1992.

Gérard Verbeke, 'Ort und Raum nach Aristoteles und Simplikios, eine philosophische Topologie', in J. Irmscher and R. Miller, eds, *Aristoteles als Wissenschaftstheoretiker*, Berlin 1983, 113-22.

C. Wildberg, *John Philoponus' Criticism of Aristotle's Theory of Ether*, Berlin and New York, 1988.

Wolfgang Wieland, 'Zur Raumtheorie des Johannes Philoponus', in *Festschrift J. Klein*, Gottingen 1967, 114-35.

Time

Julia Annas, 'Aristotle, number and time', *Philosophical Quarterly* 25, 1975, 97-113.

A.H. Armstrong, 'Eternity, life and movement in Plotinus' accounts of *nous*', *Le Néoplatonisme* (Report of the International Conference held at Royamont, 9-13 June 1969), Paris 1971, 67-74.

W. Beierwaltes, *Plotin über Ewigkeit und Zeit*, Frankfurt 1967 (text of *Enneads* 3.7, with German translation, introduction and commentary).

W. Beierwaltes, *Proklos. Grundzüge seiner Metaphysik*, 2nd edition, Frankfurt 1979.

Rémi Braque, *Du Temps chez Platon et Aristote*, Paris 1982.

P.F. Conen, *Die Zeittheorie des Aristoteles*, Munich 1964.

Pierre Duhem, *La Système du monde*, 8 vol. 1, repr. Paris 1954; vol. 2 repr. Paris 1974 (1st ed. Paris 1913-17).

Marie-Claire Galperine, 'Le temps intégral selon Damascius', *Les Études Philosophiques* 1980, 325-41.

Philippe Hoffmann, 'Jamblique, exégète du Pythagoricien Archytas: trois originalités d'une doctrine du temps', *Les Études Philosophiques* 1980, 307-23.

Philippe Hoffman, 'Paratasis', *Revue des Études Grecques* 96, 1983, 1-26.

Pamela Huby, 'An excerpt from Boethus of Sidon's commentary on the *Categories*?', *Classical Quarterly* 31, 1981, 398-409.

Edward Hussey, *Aristotle's Physics, Books III and IV*, Oxford 1983.

William Kneale, 'Time and eternity in theology', *Proceedings of the Aristotelian Society* 61, 1960-1, 87-108.

W. von Leyden, 'Time, number and eternity in Plato and Aristotle', *Philosophical Quarterly* 14, 1964, 35-52.

H. Meyer, *Das Corollarium de Tempore des Simplikios und die Aporien des Aristoteles zur Zeit*, Meisenheim am Glan 1969.

W. O'Neill, 'Time and eternity in Proclus', *Phronesis* 7, 1962, 161-5.

G.E.L. Owen, 'Aristotle on time', in Peter Machamer and Robert Turnbull, eds., *Motion and Time, Space and Matter: Interrelations in the History and Philosophy of Science*, Ohio State University 1976, repr. in Jonathan Barnes, Malcolm Schofield, Richard Sorabji, eds., *Articles on Aristotle*, vol. 3, London 1979.

S. Sambursky, *The Physical World of Late Antiquity*, London 1962, ch. 1.

Shmuel Sambursky and Shlomo Pines, *The Concept of Time in Late Neoplatonism*, Jerusalem 1971.

Lucas Siorvanes, *Proclus*, London (forthcoming).

Robert Sharples, in collaboration with F.W. Zimmermann, 'Alexander of Aphrodisias, *On Time*', *Phronesis* 1982.

E. Sonderegger, *Simplikios zur Zeit*, *Hypomnemata* vol. 70, Göttingen 1982.

Richard Sorabji, *Time, Creation and the Continuum: theories in antiquity and the early middle ages*, London and Ithaca N.Y. 1983.

Eleonore Stump and Norman Kretzmann, 'Eternity', *Journal of Philosophy* 78, 1981, 429-58.

T.A. Szlezák, *Pseudo-Archytas über die Kategorien*, Berlin 1972.

J.O. Urmson, tr., *Simplicius: Corollaries on Place and Time*, London and Ithaca N.Y. 1992.

John Whittaker, 'Ammonius on the Delphic E', *Classical Quarterly*, n.s. 19, 1969, 185-92.

John Whittaker, *God, Time, Being, Two Studies*, Symbolae Osloenses, supp. vol. 23, 1971, 1-66. Second study, pp. 33-66.

H.A. Wolfson, *Philo*, vols 1 and 2, Cambridge, Mass. 1947.

Appendix

The Commentators*

The 15,000 pages of the Ancient Greek Commentaries on Aristotle are the largest corpus of Ancient Greek philosophy that has not been translated into English or other modern European languages. The standard edition (*Commentaria in Aristotelem Graeca*, or *CAG*) was produced by Hermann Diels as general editor under the auspices of the Prussian Academy in Berlin. Arrangements have now been made to translate at least a large proportion of this corpus, along with some other Greek and Latin commentaries not included in the Berlin edition, and some closely related non-commentary works by the commentators.

The works are not just commentaries on Aristotle, although they are invaluable in that capacity too. One of the ways of doing philosophy between A.D. 200 and 600, when the most important items were produced, was by writing commentaries. The works therefore represent the thought of the Peripatetic and Neoplatonist schools, as well as expounding Aristotle. Furthermore, they embed fragments from all periods of Ancient Greek philosophical thought: this is how many of the Presocratic fragments were assembled, for example. Thus they provide a panorama of every period of Ancient Greek philosophy.

The philosophy of the period from A.D. 200 to 600 has not yet been intensively explored by philosophers in English-speaking countries, yet it is full of interest for physics, metaphysics, logic, psychology, ethics and religion. The contrast with the study of the Presocratics is striking. Initially the incomplete Presocratic fragments might well have seemed less promising, but their interest is now widely known, thanks to the philological and philosophical effort that has been concentrated upon them. The incomparably vaster corpus which preserved so many of those fragments offers at least as much interest, but is still relatively little known.

The commentaries represent a missing link in the history of philosophy: the Latin-speaking Middle Ages obtained their knowledge of Aristotle at least partly through the medium of the commentaries. Without an appreciation of this, mediaeval interpretations of Aristotle will not be understood. Again, the ancient commentaries are the unsuspected source of ideas which have been thought, wrongly, to originate in the later mediaeval

* Reprinted from the Editor's General Introduction to the series in Christian Wildberg, *Philoponus Against Aristotle on the Eternity of the World*, London and Ithaca N.Y., 1987.

period. It has been supposed, for example, that Bonaventure in the thirteenth century invented the ingenious arguments based on the concept of infinity which attempt to prove the Christian view that the universe had a beginning. In fact, Bonaventure is merely repeating arguments devised by the commentator Philoponus 700 years earlier and preserved in the meantime by the Arabs. Bonaventure even uses Philoponus' original examples. Again, the introduction of impetus theory into dynamics, which has been called a scientific revolution, has been held to be an independent invention of the Latin West, even if it was earlier discovered by the Arabs or their predecessors. But recent work has traced a plausible route by which it could have passed from Philoponus, via the Arabs, to the West.

The new availability of the commentaries in the sixteenth century, thanks to printing and to fresh Latin translations, helped to fuel the Renaissance break from Aristotelian science. For the commentators record not only Aristotle's theories, but also rival ones, while Philoponus as a Christian devises rival theories of his own and accordingly is mentioned in Galileo's early works more frequently than Plato.[1]

It is not only for their philosophy that the works are of interest. Historians will find information about the history of schools, their methods of teaching and writing and the practices of an oral tradition.[2] Linguists will find the indexes and translations an aid for studying the development of word meanings, almost wholly uncharted in Liddell and Scott's *Lexicon*, and for checking shifts in grammatical usage.

Given the wide range of interests to which the volumes will appeal, the aim is to produce readable translations, and to avoid so far as possible presupposing any knowledge of Greek. Footnotes will explain points of meaning, give cross-references to other works, and suggest alternative interpretations of the text where the translator does not have a clear preference. The introduction to each volume will include an explanation why the work was chosen for translation: none will be chosen simply because it is there. Two of the Greek texts are currently being re-edited –

[1] See Fritz Zimmermann, 'Philoponus' impetus theory in the Arabic tradition'; Charles Schmitt, 'Philoponus' commentary on Aristotle's *Physics* in the sixteenth century', and Richard Sorabji, 'John Philoponus', in Richard Sorabji (ed.), *Philoponus and the Rejection of Aristotelian Science* (London and Ithaca, N.Y. 1987).

[2] See e.g. Karl Praechter, 'Die griechischen Aristoteleskommentare', *Byzantinische Zeitschrift* 18 (1909), 516-38 (translated into English in R. Sorabji (ed.), *Aristotle Transformed: the ancient commentators and their influence* (London and Ithaca, N.Y. 1990)); M. Plezia, *de Commentariis Isagogicis* (Cracow 1947); M. Richard, '*Apo Phônês*', *Byzantion* 20 (1950), 191-222; É. Evrard, *L'Ecole d'Olympiodore et la composition du commentaire à la physique de Jean Philopon*, Diss. (Liège 1957); L.G. Westerink, *Anonymous Prolegomena to Platonic Philosophy* (Amsterdam 1962) (new revised edition, translated into French, Collection Budé; part of the revised introduction, in English, is included in *Aristotle Transformed*); A.-J. Festugière, 'Modes de composition des commentaires de Proclus', *Museum Helveticum* 20 (1963), 77-100, repr. in his *Études* (1971), 551-74; P. Hadot, 'Les divisions des parties de la philosophie dans l'antiquité', *Museum Helveticum* 36 (1979), 201-23; I. Hadot, 'La division néoplatonicienne des écrits d'Aristote', in J. Wiesner (ed.), *Aristoteles Werk und Wirkung* (Paul Moraux gewidmet), vol. 2 (Berlin 1986); I. Hadot, 'Les introductions aux commentaires exégétiques chez les auteurs néoplatoniciens et les auteurs chrétiens', in M. Tardieu (ed.), *Les règles de l'interprétation* (Paris 1987), 99-119. These topics are treated, and a bibliography supplied, in *Aristotle Transformed*.

those of Simplicius *in Physica* and *in de Caelo* – and new readings will be exploited by translators as they become available. Each volume will also contain a list of proposed emendations to the standard text. Indexes will be of more uniform extent as between volumes than is the case with the Berlin edition, and there will be three of them: an English-Greek glossary, a Greek-English index, and a subject index.

The commentaries fall into three main groups. The first group is by authors in the Aristotelian tradition up to the fourth century A.D. This includes the earliest extant commentary, that by Aspasius in the first half of the second century A.D. on the *Nicomachean Ethics*. The anonymous commentary on Books 2, 3, 4 and 5 of the *Nicomachean Ethics*, in *CAG* vol. 20, is derived from Adrastus, a generation later.[3] The commentaries by Alexander of Aphrodisias (appointed to his chair between A.D. 198 and 209) represent the fullest flowering of the Aristotelian tradition. To his successors Alexander was The Commentator *par excellence*. To give but one example (not from a commentary) of his skill at defending and elaborating Aristotle's views, one might refer to his defence of Aristotle's claim that space is finite against the objection that an edge of space is conceptually problematic.[4] Themistius (*fl.* late 340s to 384 or 385) saw himself as the inventor of paraphrase, wrongly thinking that the job of commentary was completed.[5] In fact, the Neoplatonists were to introduce new dimensions into commentary. Themistius' own relation to the Neoplatonist as opposed to the Aristotelian tradition is a matter of controversy,[6] but it would be agreed that his commentaries show far less bias than the full-blown Neoplatonist ones. They are also far more informative than the designation 'paraphrase' might suggest, and it has been estimated that Philoponus' *Physics* commentary draws silently on Themistius six hundred times.[7] The pseudo-Alexandrian commentary on *Metaphysics* 6–14, of unknown authorship, has been placed by some in the same group of commentaries as being earlier than the fifth century.[8]

[3] Anthony Kenny, *The Aristotelian Ethics* (Oxford 1978), 37, n.3; Paul Moraux, *Der Aristotelismus bei den Griechen,* vol. 2 (Berlin 1984), 323-30.

[4] Alexander, *Quaestiones* 3.12, discussed in my *Matter, Space and Motion* (London and Ithaca, N.Y. 1988). For Alexander see R.W. Sharples, 'Alexander of Aphrodisias: scholasticism and innovation', in W. Haase (ed.), *Aufstieg und Niedergang der römischen Welt*, part 2 *Principat*, vol. 36.2, *Philosophie und Wissenschaften* (1987).

[5] Themistius *in An. Post.* 1,2-12. See H.J. Blumenthal, 'Photius on Themistius (Cod.74): did Themistius write commentaries on Aristotle?', *Hermes* 107 (1979), 168-82.

[6] For different views, see H.J. Blumenthal, 'Themistius, the last Peripatetic commentator on Aristotle?', in Glen W. Bowersock, Walter Burkert, Michael C.J. Putnam, *Arktouros*, Hellenic Studies Presented to Bernard M.W. Knox (Berlin and N.Y., 1979), 391-400; E.P. Mahoney, 'Themistius and the agent intellect in James of Viterbo and other thirteenth-century philosophers: (Saint Thomas Aquinas, Siger of Brabant and Henry Bate)', *Augustiniana* 23 (1973), 422-67, at 428-31; id., 'Neoplatonism, the Greek commentators and Renaissance Aristotelianism', in D.J. O'Meara (ed.), *Neoplatonism and Christian Thought* (Albany N.Y. 1982), 169-77 and 264-82, esp. n. 1, 264-6; Robert Todd, introduction to translation of Themistius *in DA* 3.4-8, in *Two Greek Aristotelian Commentators on the Intellect*, trans. Frederick M. Schroeder and Robert B. Todd (Toronto 1990).

[7] H. Vitelli, *CAG* 17, p. 992, s.v. Themistius.

[8] The similarities to Syrianus (died *c.*437) have suggested to some that it predates Syrianus (most recently Leonardo Tarán, review of Paul Moraux, *Der Aristotelismus*,

By far the largest group of extant commentaries is that of the Neoplatonists up to the sixth century A.D. Nearly all the major Neoplatonists, apart from Plotinus (the founder of Neoplatonism), wrote commentaries on Aristotle, although those of Iamblichus (*c.* 250 – *c.* 325) survive only in fragments, and those of three Athenians, Plutarchus (died 432), his pupil Proclus (410 – 485) and the Athenian Damascius (*c.* 462 – after 538), are lost.[9] As a result of these losses, most of the extant Neoplatonist commentaries come from the late fifth and the sixth centuries and a good proportion from Alexandria. There are commentaries by Plotinus' disciple and editor Porphyry (232 – 309), by Iamblichus' pupil Dexippus (*c.* 330), by Proclus' teacher Syrianus (died *c.* 437), by Proclus' pupil Ammonius (435/445 – 517/526), by Ammonius' three pupils Philoponus (*c.* 490 to 570s), Simplicius (wrote after 532, probably after 538) and Asclepius (sixth century), by Ammonius' next but one successor Olympiodorus (495/505 – after 565), by Elias (*fl.* 541?), by David (second half of the sixth century, or beginning of the seventh) and by Stephanus (took the chair in Constantinople *c.* 610). Further, a commentary on the *Nicomachean Ethics* has been ascribed to Heliodorus of Prusa, an unknown pre-fourteenth-century figure, and there is a commentary by Simplicius' colleague Priscian of Lydia on Aristotle's successor Theophrastus. Of these commentators some of the last were Christians (Philoponus, Elias, David and Stephanus), but they were Christians writing in the Neoplatonist tradition, as was also Boethius who produced a number of commentaries in Latin before his death in 525 or 526.

The third group comes from a much later period in Byzantium. The Berlin edition includes only three out of more than a dozen commentators described in Hunger's *Byzantinisches Handbuch*.[10] The two most important are Eustratius (1050/1060 – *c.* 1120), and Michael of Ephesus. It has been suggested that these two belong to a circle organised by the princess Anna Comnena in the twelfth century, and accordingly the completion of Michael's commentaries has been redated from 1040 to 1138.[11] His commentaries include areas where gaps had been left. Not all of these gap-fillers are extant, but we have commentaries on the neglected biological works, on the *Sophistici Elenchi*, and a small fragment of one on the *Politics*. The lost *Rhetoric* commentary had a few antecedents, but the *Rhetoric* too had been comparatively neglected. Another product of this

---

vol. 1, in *Gnomon* 46 (1981), 721-50 at 750), to others that it draws on him (most recently P. Thillet, in the Budé edition of Alexander *de Fato*, p. lvii). Praechter ascribed it to Michael of Ephesus (eleventh or twelfth century), in his review of *CAG* 22.2, in *Göttingische Gelehrte Anzeiger* 168 (1906), 861-907.

[9] The Iamblichus fragments are collected in Greek by Bent Dalsgaard Larsen, *Jamblique de Chalcis, Exégète et Philosophe* (Aarhus 1972), vol.2. Most are taken from Simplicius, and will accordingly be translated in due course. The evidence on Damascius' commentaries is given in L.G. Westerink, *The Greek Commentaries on Plato's Phaedo*, vol.2, Damascius (Amsterdam 1977), 11-12; on Proclus' in L.G. Westerink, *Anonymous Prolegomena to Platonic Philosophy* (Amsterdam 1962), xii, n.22; on Plutarchus' in H.M. Blumenthal, 'Neoplatonic elements in the de Anima commentaries', *Phronesis* 21 (1976), 75.

[10] Herbert Hunger, *Die hochsprachliche profane Literatur der Byzantiner*, vol.1 (= *Byzantinisches Handbuch*, part 5, vol.1) (Munich 1978), 25-41. See also B.N. Tatakis, *La Philosophie Byzantine* (Paris 1949).

[11] R. Browning, 'An unpublished funeral oration on Anna Comnena', *Proceedings of the Cambridge Philological Society* n.s. 8 (1962), 1-12, esp. 6-7.

period may have been the composite commentary on the *Nicomachean Ethics* (*CAG* 20) by various hands, including Eustratius and Michael, along with some earlier commentators, and an improvisation for Book 7. Whereas Michael follows Alexander and the conventional Aristotelian tradition, Eustratius' commentary introduces Platonist, Christian and anti-Islamic elements.[12]

The composite commentary was to be translated into Latin in the next century by Robert Grosseteste in England. But Latin translations of various logical commentaries were made from the Greek still earlier by James of Venice (*fl. c.* 1130), a contemporary of Michael of Ephesus, who may have known him in Constantinople. And later in that century other commentaries and works by commentators were being translated from Arabic versions by Gerard of Cremona (died 1187).[13] So the twelfth century resumed the transmission which had been interrupted at Boethius' death in the sixth century.

The Neoplatonist commentaries of the main group were initiated by Porphyry. His master Plotinus had discussed Aristotle, but in a very independent way, devoting three whole treatises (*Enneads* 6.1–3) to attacking Aristotle's classification of the things in the universe into categories. These categories took no account of Plato's world of Ideas, were inferior to Plato's classifications in the *Sophist* and could anyhow be collapsed, some of them into others. Porphyry replied that Aristotle's categories could apply perfectly well to the world of intelligibles and he took them as in general defensible.[14] He wrote two commentaries on the *Categories*, one lost, and an introduction to it, the *Isagôgê*, as well as commentaries, now lost, on a number of other Aristotelian works. This proved decisive in making Aristotle a necessary subject for Neoplatonist lectures and commentary. Proclus, who was an exceptionally quick student, is said to have taken two years over his Aristotle studies, which were called

[12] R. Browning, op. cit. H.D.P. Mercken, *The Greek Commentaries of the Nicomachean Ethics of Aristotle in the Latin Translation of Grosseteste, Corpus Latinum Commentariorum in Aristotelem Graecorum* VI 1 (Leiden 1973), ch.1, 'The compilation of Greek commentaries on Aristotle's Nicomachean Ethics'. Sten Ebbesen, 'Anonymi Aurelianensis I Commentarium in *Sophisticos Elenchos*', *Cahiers de l'Institut Moyen Age Grecque et Latin* 34 (1979), 'Boethius, Jacobus Veneticus, Michael Ephesius and "Alexander" ', pp. v-xiii; id., *Commentators and Commentaries on Aristotle's Sophistici Elenchi*, 3 parts, *Corpus Latinum Commentariorum in Aristotelem Graecorum*, vol. 7 (Leiden 1981); A. Preus, *Aristotle and Michael of Ephesus on the Movement and Progression of Animals* (Hildesheim 1981), introduction.

[13] For Grosseteste, see Mercken as in n. 12. For James of Venice, see Ebbesen as in n. 12, and L. Minio-Paluello, 'Jacobus Veneticus Grecus', *Traditio* 8 (1952), 265-304; id., 'Giacomo Veneto e l'Aristotelismo Latino', in Pertusi (ed.), *Venezia e l'Oriente fra tardo Medioevo e Rinascimento* (Florence 1966), 53-74, both reprinted in his *Opuscula* (1972). For Gerard of Cremona, see M. Steinschneider, *Die europäischen Übersetzungen aus dem arabischen bis Mitte des 17. Jahrhunderts* (repr. Graz 1956); E. Gilson, *History of Christian Philosophy in the Middle Ages* (London 1955), 235-6 and more generally 181-246. For the translators in general, see Bernard G. Dod, 'Aristoteles Latinus', in N. Kretzmann, A. Kenny, J. Pinborg (eds). *The Cambridge History of Latin Medieval Philosophy* (Cambridge 1982).

[14] See P. Hadot, 'L'harmonie des philosophies de Plotin et d'Aristote selon Porphyre dans le commentaire de Dexippe sur les Catégories', in *Plotino e il neoplatonismo in Oriente e in Occidente* (Rome 1974), 31-47; A.C. Lloyd, 'Neoplatonic logic and Aristotelian logic', *Phronesis* 1 (1955-6), 58-79 and 146-60.

the Lesser Mysteries, and which preceded the Greater Mysteries of Plato.[15] By the time of Ammonius, the commentaries reflect a teaching curriculum which begins with Porphyry's *Isagôgê* and Aristotle's *Categories*, and is explicitly said to have as its final goal a (mystical) ascent to the supreme Neoplatonist deity, the One.[16] The curriculum would have progressed from Aristotle to Plato, and would have culminated in Plato's *Timaeus* and *Parmenides*. The latter was read as being about the One, and both works were established in this place in the curriculum at least by the time of Iamblichus, if not earlier.[17]

Before Porphyry, it had been undecided how far a Platonist should accept Aristotle's scheme of categories. But now the proposition began to gain force that there was a harmony between Plato and Aristotle on most things.[18] Not for the only time in the history of philosophy, a perfectly crazy proposition proved philosophically fruitful. The views of Plato and of Aristotle had both to be transmuted into a new Neoplatonist philosophy in order to exhibit the supposed harmony. Iamblichus denied that Aristotle contradicted Plato on the theory of Ideas.[19] This was too much for Syrianus and his pupil Proclus. While accepting harmony in many areas,[20] they could see that there was disagreement on this issue and also on the issue of whether God was causally responsible for the existence of the ordered physical cosmos, which Aristotle denied. But even on these issues, Proclus' pupil Ammonius was to claim harmony, and, though the debate was not clear cut,[21] his claim was on the whole to prevail. Aristotle, he maintained, accepted Plato's Ideas,[22] at least in the form of principles (*logoi*) in the divine intellect, and these principles were in turn causally responsible for the beginningless existence of the physical universe. Ammonius wrote a whole book to show that

[15] Marinus, *Life of Proclus* ch.13, 157,41 (Boissonade).

[16] The introductions to the *Isagôgê* by Ammonius, Elias and David, and to the *Categories* by Ammonius, Simplicius, Philoponus, Olympiodorus and Elias are discussed by L.G. Westerink, *Anonymous Prolegomena* and I. Hadot, 'Les Introductions', see n. 2 above.

[17] Proclus *in Alcibiadem 1* p.11 (Creuzer); Westerink, *Anonymous Prolegomena*, ch. 26, 12f. For the Neoplatonist curriculum see Westerink, Festugière, P. Hadot and I. Hadot in n. 2.

[18] See e.g. P. Hadot (1974), as in n. 14 above; H.J. Blumenthal, 'Neoplatonic elements in the de Anima commentaries', *Phronesis* 21 (1976), 64-87; H.A. Davidson, 'The principle that a finite body can contain only finite power', in S. Stein and R. Loewe (eds), *Studies in Jewish Religious and Intellectual History presented to A. Altmann* (Alabama 1979), 75-92; Carlos Steel, 'Proclus et Aristote', Proceedings of the Congrès Proclus held in Paris 1985, J. Pépin and H.D. Saffrey (eds), *Proclus, lecteur et interprète des anciens* (Paris 1987), 213-25; Koenraad Verrycken, *God en Wereld in de Wijsbegeerte van Ioannes Philoponus*, Ph.D. Diss. (Louvain 1985).

[19] Iamblichus ap. Elian *in Cat.* 123,1-3.

[20] Syrianus *in Metaph.* 80,4-7; Proclus *in Tim.* 1.6,21-7,16.

[21] Asclepius sometimes accepts Syranius' interpretation (*in Metaph.* 433,9-436,6); which is, however, qualified, since Syrianus thinks Aristotle is really committed willy-nilly to much of Plato's view (*in Metaph.* 117,25-118,11; ap. Asclepium *in Metaph.* 433,16; 450,22); Philoponus repents of his early claim that Plato is not the target of Aristotle's attack, and accepts that Plato is rightly attacked for treating ideas as independent entities outside the divine Intellect (*in DA* 37,18-31; *in Phys.* 225,4-226,11; *contra Procl.* 26,24-32,13; *in An. Post.* 242,14–243,25).

[22] Asclepius *in Metaph.* from the voice of (i.e. from the lectures of) Ammonius 69,17-21; 71,28; cf. Zacharias *Ammonius, Patrologia Graeca* vol. 85, col. 952 (Colonna).

Aristotle's God was thus an efficient cause, and though the book is lost, some of its principal arguments are preserved by Simplicius.[23] This tradition helped to make it possible for Aquinas to claim Aristotle's God as a Creator, albeit not in the sense of giving the universe a beginning, but in the sense of being causally responsible for its beginningless existence.[24] Thus what started as a desire to harmonise Aristotle with Plato finished by making Aristotle safe for Christianity. In Simplicius, who goes further than anyone,[25] it is a formally stated duty of the commentator to display the harmony of Plato and Aristotle in most things.[26] Philoponus, who with his independent mind had thought better of his earlier belief in harmony, is castigated by Simplicius for neglecting this duty.[27]

The idea of harmony was extended beyond Plato and Aristotle to Plato and the Presocratics. Plato's pupils Speusippus and Xenocrates saw Plato as being in the Pythagorean tradition.[28] From the third to first centuries B.C., pseudo-Pythagorean writings present Platonic and Aristotelian doctrines as if they were the ideas of Pythagoras and his pupils,[29] and these forgeries were later taken by the Neoplatonists as genuine. Plotinus saw the Presocratics as precursors of his own views,[30] but Iamblichus went far beyond him by writing ten volumes on Pythagorean philosophy.[31] Thereafter Proclus sought to unify the whole of Greek philosophy by presenting it as a continuous clarification of divine revelation,[32] and Simplicius argued for the same general unity in order to rebut Christian charges of contradictions in pagan philosophy.[33]

Later Neoplatonist commentaries tend to reflect their origin in a teaching curriculum:[34] from the time of Philoponus, the discussion is often divided up into lectures, which are subdivided into studies of doctrine and of text. A general account of Aristotle's philosophy is prefixed to the *Categories* commentaries and divided, according to a formula of Proclus,[35] into ten questions. It is here that commentators explain the eventual purpose of studying Aristotle (ascent to the One) and state (if they do) the requirement of displaying the harmony of Plato and Aristotle. After the ten-point introduction to Aristotle, the *Categories* is given a six-point introduction, whose antecedents go back earlier than Neoplatonism, and which requires

[23] Simplicius *in Phys.* 1361,11-1363,12. See H.A. Davidson; Carlos Steel; Koenraad Verrycken in n.18 above.

[24] See Richard Sorabji, *Matter, Space and Motion* (London and Ithaca N.Y. 1988), ch. 15.

[25] See e.g. H.J. Blumenthal in n. 18 above.

[26] Simplicius *in Cat.* 7,23-32.

[27] Simplicius *in Cael.* 84,11-14; 159,2-9. On Philoponus' *volte face* see n. 21 above.

[28] See e.g. Walter Burkert, *Weisheit und Wissenschaft* (Nürnberg 1962), translated as *Lore and Science in Ancient Pythagoreanism* (Cambridge Mass. 1972), 83-96.

[29] See Holger Thesleff, *An Introduction to the Pythagorean writings of the Hellenistic Period* (Åbo 1961); Thomas Alexander Szlezák, *Pseudo-Archytas über die Kategorien*, Peripatoi vol. 4 (Berlin and New York 1972).

[30] Plotinus e.g. 4.8.1; 5.1.8 (10-27); 5.1.9.

[31] See Dominic O'Meara, *Pythagoras Revived: Mathematics and Philosophy in late Antiquity* (Oxford 1989).

[32] See Christian Guérard, 'Parménide d'Elée selon les Néoplatoniciens', forthcoming.

[33] Simplicius *in Phys.* 28,32-29,5; 640,12-18. Such thinkers as Epicurus and the Sceptics, however, were not subject to harmonisation.

[34] See the literature in n. 2 above.

[35] ap. Elian *in Cat.* 107,24-6.

the commentator to find a unitary theme or scope (*skopos*) for the treatise. The arrangements for late commentaries on Plato are similar. Since the Plato commentaries form part of a single curriculum they should be studied alongside those on Aristotle. Here the situation is easier, not only because the extant corpus is very much smaller, but also because it has been comparatively well served by French and English translators.[36]

Given the theological motive of the curriculum and the pressure to harmonise Plato with Aristotle, it can be seen how these commentaries are a major source for Neoplatonist ideas. This in turn means that it is not safe to extract from them the fragments of the Presocratics, or of other authors, without making allowance for the Neoplatonist background against which the fragments were originally selected for discussion. For different reasons, analogous warnings apply to fragments preserved by the pre-Neoplatonist commentator Alexander.[37] It will be another advantage of the present translations that they will make it easier to check the distorting effect of a commentator's background.

Although the Neoplatonist commentators conflate the views of Aristotle with those of Neoplatonism, Philoponus alludes to a certain convention when he quotes Plutarchus expressing disapproval of Alexander for expounding his own philosophical doctrines in a commentary on Aristotle.[38] But this does not stop Philoponus from later inserting into his own commentaries on the *Physics* and *Meteorology* his arguments in favour of the Christian view of Creation. Of course, the commentators also wrote independent works of their own, in which their views are expressed independently of the exegesis of Aristotle. Some of these independent works will be included in the present series of translations.

The distorting Neoplatonist context does not prevent the commentaries from being incomparable guides to Aristotle. The introductions to Aristotle's philosophy insist that commentators must have a minutely detailed knowledge of the entire Aristotelian corpus, and this they certainly have. Commentators are also enjoined neither to accept nor reject what Aristotle says too readily, but to consider it in depth and without partiality. The commentaries draw one's attention to hundreds of phrases, sentences and ideas in Aristotle, which one could easily have passed over, however often one read him. The scholar who makes the right allowance for the distorting context will learn far more about Aristotle than he would be likely to on his own.

The relations of Neoplatonist commentators to the Christians were subtle. Porphyry wrote a treatise explicitly against the Christians in 15 books, but an order to burn it was issued in 448, and later Neoplatonists

[36] English: Calcidius *in Tim.* (parts by van Winden; den Boeft); Iamblichus fragments (Dillon); Proclus *in Tim.* (Thomas Taylor); Proclus *in Parm.* (Dillon); Proclus *in Parm.*, end of 7th book, from the Latin (Klibansky, Labowsky, Anscombe); Proclus *in Alcib. 1* (O'Neill); Olympiodorus and Damascius *in Phaedonem* (Westerink); Damascius *in Philebum* (Westerink); *Anonymous Prolegomena to Platonic Philosophy* (Westerink). See also extracts in Thomas Taylor, *The Works of Plato*, 5 vols. (1804). French: Proclus *in Tim.* and *in Rempublicam* (Festugière); *in Parm.* (Chaignet); Anon. *in Parm.* (P. Hadot); Damascius *in Parm.* (Chaignet).

[37] For Alexander's treatment of the Stoics, see Robert B. Todd, *Alexander of Aphrodisias on Stoic Physics* (Leiden 1976), 24-9.

[38] Philoponus *in DA* 21.20-3.

were more circumspect. Among the last commentators in the main group, we have noted several Christians. Of these the most important were Boethius and Philoponus. It was Boethius' programme to transmit Greek learning to Latin-speakers. By the time of his premature death by execution, he had provided Latin translations of Aristotle's logical works, together with commentaries in Latin but in the Neoplatonist style on Porphyry's *Isagôgê* and on Aristotle's *Categories* and *de Interpretatione*, and interpretations of the *Prior* and *Posterior Analytics, Topics* and *Sophistici Elenchi*. The interruption of his work meant that knowledge of Aristotle among Latin-speakers was confined for many centuries to the logical works. Philoponus is important both for his proofs of the Creation and for his progressive replacement of Aristotelian science with rival theories, which were taken up at first by the Arabs and came fully into their own in the West only in the sixteenth century.

Recent work has rejected the idea that in Alexandria the Neoplatonists compromised with Christian monotheism by collapsing the distinction between their two highest deities, the One and the Intellect. Simplicius (who left Alexandria for Athens) and the Alexandrians Ammonius and Asclepius appear to have acknowledged their beliefs quite openly, as later did the Alexandrian Olympiodorus, despite the presence of Christian students in their classes.[39]

The teaching of Simplicius in Athens and that of the whole pagan Neoplatonist school there was stopped by the Christian Emperor Justinian in 529. This was the very year in which the Christian Philoponus in Alexandria issued his proofs of Creation against the earlier Athenian Neoplatonist Proclus. Archaeological evidence has been offered that, after their temporary stay in Ctesiphon (in present-day Iraq), the Athenian Neoplatonists did not return to their house in Athens, and further evidence has been offered that Simplicius went to Ḥarrān (Carrhae), in present-day Turkey near the Iraq border.[40] Wherever he went, his commentaries are a treasure house of information about the preceding thousand years of Greek philosophy, information which he painstakingly recorded after the closure in Athens, and which would otherwise have been lost. He had every reason to feel bitter about Christianity, and in fact he sees it and Philoponus, its representative, as irreverent. They deny the divinity of the heavens and prefer the physical relics of dead martyrs.[41] His own commentaries by

[39] For Simplicius, see I. Hadot, *Le Problème du Néoplatonisme Alexandrin: Hiéroclès et Simplicius* (Paris 1978); for Ammonius and Asclepius, Koenraad Verrycken, *God en Wereld in de Wijsbegeerte van Ioannes Philoponus*, Ph.D. Diss. (Louvain 1985); for Olympiodorus, L.G. Westerink, *Anonymous Prolegomena to Platonic Philosophy* (Amsterdam 1962).

[40] Alison Frantz, 'Pagan philosophers in Christian Athens', *Proceedings of the American Philosophical Society* 119 (1975), 29-38; M. Tardieu, 'Témoins orientaux du *Premier Alcibiade* à Ḥarrān et à Nag 'Hammādi', *Journal Asiatique* 274 (1986); id., 'Les calendriers en usage à Ḥarrān d'après les sources arabes et le commentaire de Simplicius à la *Physique* d'Aristote', in I. Hadot (ed.), *Simplicius, sa vie, son oeuvre, sa survie* (Berlin 1987), 40-57; id., *Coutumes nautiques mésopotamiennes chez Simplicius*, in preparation. The opposing view that Simplicius returned to Athens is most fully argued by Alan Cameron, 'The last days of the Academy at Athens', *Proceedings of the Cambridge Philological Society* 195, n.s. 15 (1969), 7-29.

[41] Simplicius *in Cael.* 26,4-7; 70,16-18; 90,1-18; 370,29-371,4. See on his whole attitude Philippe Hoffmann, 'Simplicius' polemics', in Richard Sorabji (ed.), *Philoponus and the Rejection of Aristotelian Science* (London and Ithaca, N.Y. 1987).

contrast culminate in devout prayers.

Two collections of articles by various hands have been published, to make the work of the commentators better known. The first is devoted to Philoponus;[42] the second is about the commentators in general, and goes into greater detail on some of the issues briefly mentioned here.[43]

[42] Richard Sorabji (ed.), *Philoponus and the Rejection of Aristotelian Science* (London and Ithaca, N.Y. 1987).

[43] Richard Sorabji (ed.), *Aristotle Transformed: the ancient commentators and their influence* (London and Ithaca, N.Y. 1990). The lists of texts and previous translations of the commentaries included in Wildberg, *Philoponus Against Aristotle on the Eternity of the World* (pp.12ff.) are not included here. The list of translations should be augmented by: F.L.S. Bridgman, Heliodorus (?) in *Ethica Nicomachea*, London 1807.

I am grateful for comments to Henry Blumenthal, Victor Caston, I. Hadot, Paul Mercken, Alain Segonds, Robert Sharples, Robert Todd, L.G. Westerink and Christian Wildberg.

English–Greek Glossary

abstract (v.): *aphairein*
abstract (number): *monadikos*
abstraction: *aphairesis*
account: *logos*
actuality: *energeia*, *entelekheia*
actually: *energeiāi*
ad infinitum: *eis apeiron*, *ep' apeiron*
adjoining: *ephexês*
aether: *aithêr*
to be affected: *paskhein*
affection: *pathos*
after: *husteros*
akin: *sungenês*
already: *êdê*
alteration: *alloiôsis*
ambiguous: *homônumos*
antecedent: *hêgoumenon*
appear: *phainesthai*
applied (number): *methektos*
argument: *epikheirêma*
articulate: *diarthroun*
as such: *kath' hauto*
assumption: *axiôma*
attribute: *pathos*, *to sumbebêkos*

to be: *einai*
become: *ginesthai*
becoming: *genesis*
before: *proteros*
being: *to einai*, *ousia*
bodily: *sômatoeides*
body: *sôma*
breadth: *platos*
bulk: *onkos*

capacity: *dunamis*
cause: *aitia*, *aition*
change (v.): *kinein*, *kineisthai*, *metaballein* (see n. 1 to the Translation for other words for change)
change: *kinêsis*, *metabolê* (see nn. 1 and 163 to the Translation)
  change of place: *kinêsis kata topon*
  qualitative change: *kinêsis kata poion*
coalesce: *sumphuesthai*
commentator: *exêgêtês*, *hupomnêmatistos*
common: *koinos*
complete: *teleios*
conceive: *epinoein*
concept: *ennoia*
conceptually: *epinoiāi*
conclude: *sumperainesthai*, *sunagein*
  drawing to a conclusion: *sunagôgê*
conclusion: *sumperasma*
condition: *hexis*
consequent: *akolouthia*, *hepomenon*
contain: *periekhein*
container: *periekhon*
contingently: *kata sumbebêkos*
continuous: *sunekhês*
conversion: *antistrophê*
criterion: *tekmêrion*
cyclic motion: *anakuklêsis*, *kuklophoria*, *periphora*

definition: *horismos*
demonstration: *apodeixis*
demonstrative: *apodeiktikos*
depth: *bathos*
difference: *diaphora*
differentia: *diaphora*
difficulty: *aporia*
dimension: *diastêma*, *diastasis*
discrete: *diêirêmenos*
displacement: *antimetastasis*
divine: *theios*
divisible: *diairetos*
division: *diairesis*

duration: *paratasis*

earlier: *proteros*
empty: *kenos*
end: *telos*
enumerate: *arithmein*
enumerated: *arithmoumenos*
essence: *ousia*
eternal: *aïdios*, *aiônios*
eternity: *aiôn*
everlasting: *aïdios*
excellence: *aretê*
exist: *huphistanai*
existence: *huparxis*, *hupostasis*
explanation: *aitia*
extended: *diestôs*
extension: *diastasis*, *paratasis*

figure (of a syllogism): *skhêma*
fixed: *aplanês*
flow: *rhoê*
by force: *biāi*
form: *eidos*
future: *mellon*

genus: *genos*
great: *megas*
growth: *auxêsis*

heavenly: *ouranios*
heavens: *ouranos*
here: *entha*, *enthade*
holding together: *sustatikos*

immaterial: *aülos*
immobile: *akinêtos*
implausible: *apithanos*
impulse: *hormê*
incorporeal: *asômatos*
indivisible: *adiairetos*
infinite: *apeiros*
inseparable: *akhôristos*
instantaneous: *akhronos*
intelligence: *nous*
intermediate: *mesos*
interval: *diastêma*

knowledge: *epistêmê*

later: *husteros*
length: *mêkos*
limit: *peras*
limited: *peperasmenos*
local: *topikos*
locomotion: *phora*

magnitude: *megethos*
manuscript: *antigraphon*, *graphê*
material: *hulikos*
matter: *hulê*
measure (n.): *metron*
measure (v.): *metrein*
motion: *phora*, *kinêsis*
move (v.) *kinein*, *kineisthai*, *pheresthai* (see n. 1 to the Translation for other words for motion)
moveable: *metaphorêtos*

natural: *phusikos*
nature: *phusis*
nontechnical: *exôterikos*
now: *nun*
number: *arithmos*

order: *taxis*

paradigm: *paradeigma*
part: *meros*
participate: *metalambanein*
participated: *methektos*
participation: *methexis*
passing away: *phthora*
past: *parelêluthos*
permeate: *khôrein dia*
place: *topos*
plane: *epipedos*
point (mathematical): *sêmeion*
portion: *morion*
position: *thesis*
postulate: *axiôma*
potentially: *dunamei*
power: *dunamis*
pre-eminently: *proêgoumenôs*
premise: *protasis*
  major premise: *meizôn protasis*
  minor premise: *elattôn protasis*
present: *enestôs*
primarily: *prôtôs*
privation: *sterêsis*
problem: *aporia*, *problêma*
proceed: *proienai*

procession: *proödos*
proof: *deixis*
proper: *idios*, *oikeios*

quality: *poion*
quantity: *poson*

rational: *logikos*, *logistikos*
reading (of text): *graphê*
receptacle: *hupodokhê*
receptive: *khôrêtikos*
related: *oikeios*
remain: *menein*
rest: *êremia*

separate: *khôristos*
shape: *morphê*, *skhêma*
sign: *sêmeion*
signify: *sêmainein*
simultaneously: *hama*
small: *mikros*
sometime: *pote*
soul: *psukhê*
sound: *hugiês*
space: *khôra*
species: *eidos*
sphere: *sphaira*
straight: *euthus*
substance: *ousia*
substrate: *hupokeimenon*
suddenly: *exaiphnês*
surface: *epiphaneia*
syllogism: *sullogismos*

technical: *akroamatikos*
temporal: *khronikos*
there: *ekei*
think: *noein*
thought: *noêsis*
time: *khronos*
totality: *to pan*
touch: *haptesthai*
transformation: *metabolê*

understand: *sunnoein*
understanding: *katalêpsis*, *katanoêsis*
undifferentiated: *adiaphoros*
unforced: *abiastos*
universe: *kosmos*, *to pan*
unlimited: *apeiros*

valid: *hugiês*
vessel: *angeion*
void: *to kenon*

well-known: *gnôrimos*

Greek–English Glossary

References are to the page and line numbers of the *CAG* edition, which appear in the margin of this translation.

abasanistôs, without examination, 762,7
abiastos, unforced, 599,23.30
adiairetos, indivisible, 531,8; 551,12; 556,20; 562,7; 568,6; 569,24; 571,18
adiaphoros, indifferent, indiscriminate, undifferentiated, identical, 521,23; 706,31.32; 721,8.15; 722,5.10; 764,34; 767,12; 771,1; 772,32
adiastatos, unextended, 600,30; 697,16; 708,33; 749,35
adunatos, impossible, 527,8; 530,1.2.14; 533,18; 534,12; 542,30; 544,29; 546,31; 548,1.3; 550,18; 552,23; 553,29.32; 554,30 *passim*
agôgê, trend, 531,16; 546,25; 759,14
agraphos, unwritten, 542,10.12; 545,23
aïdios, eternal, everlasting, 541,33; 703,31; 704,1; 718,8; 737,34; 739,14-18.22; 741,22; 742,1.17; 743,23.25.32; 745,30; 746,13; 747,21-2; 758,24; 768,8
aiôn, eternity, 704,4.5; 717,25; 718,6.8; 737,35; 742,4.16
aiônios, eternal, 702,31; 703,8; 704,3.5; 717,26.29; 739,15.17.21
aisthêtos, perceptible, 526,31; 527,13; 531,1; 532,9.18-25.35.36; 539,18; 695,23; 704,19; 753,11-13.20; 759,19; 763,4; 765,27
aithêr, aether, 528,16; 586,23.26; 587,5.10.11.14; 590,21.31; 593,12; 594,26-31
aitia, ***aition***, cause, 519,3.9.17; 522,24; 523,11.13.15.23; 527,28; 528,14.18.21.23 *passim*; *aition telikon*, perfecting cause, 519.7; 520.5; 533,7
akhôristos, inseparable, 543,25; 544,8.14.23.24; 757,14; 767,8.21; *akhôristôs*, inseparably, 544,20
akhronos, instantaneous, in no time, 753,4; *akhronôs*, instantaneously, 753,26
akinêtos, immoveable, immobile, 520,8-10; 550,34; 553,26; 566,19; 576,25; 578,9.29; 581,28.33; 582,1.5.8.34.36; 583,2.8.11.16.18.25.32; 584,10.16.18.23.27.31; 585,2.3.10.19.22.25-9; 588,3.4.23; 589,18.24; 595,6; 596,4; 707,8.19; 708,30.32; 709,1; 710,8.9; 738,32; 742,16; 743,22; 745,3.16.27.28
akolouthein, to follow, 520,22; 530,10; 534,12.18; 546,19; 557,17; 558,24.28; 559,6; 574,24; 575,27; 577,10; 567,6; 702,27; 703,1; 718,26; 722,23.27; 723,18.22; 725,1.5; 726,30; 727,3; 731,9; 732,13; 733,30; 739,3; 757,2.4; 762,8; 763,32; 765,22; *akolouthôn*, following, in accordance with, 759,21
akolouthia, consequence, 521,36; connection, relationship, dependence, 722,23; 724,9.26; 725,9; 726,3.17.28; *anaskeuastikon ex akolouthias*, denial of the consequent, 529,35; 574,21; *akolou-*

thos, following upon, consistent, 597,7; *to akolouthon*, consequence, 730,27; *akolouthoteros*, more consistent, 559,3
akroamatikos, technical, 696,1
allassein, to change (place), 568,24; to exchange (place), 581,32; 595,21; *metallassein*, to exchange, 595,8
alloios, different, 574,15
alloiôsis, qualitative change, alteration, 522,17; 711,32; 717,3; 720,25; 724,10; 737,23; 742,5; 760,25; 761,24; 762.3.14.21; 764,14.25; 768,6.10; 771,22
alloioun, to alter, 600,6-8; 705,16; 720,14.15; 724,21
ameibein, to change (place), 568,26.29; to exchange (place), 582,1; 589,8; 591,21; 592,2.4.5
amethodôs, irregularly, 521,15
anakuklêsis, revolution, cyclic motion, 701,18.19.21
anarithmêsis, reckoning up, 714,28
anarithmêtos, uncountable, 740,31
anaskeuastikos, destructive, 574,22; *anaskeuastikos ex akolouthias*, denial of the consequent, 529,35; 574,22
anekleiptos, inexhaustible, 739,27
angeion, vessel, 524,6.31; 543,11.29-33; 550,29; 551.2; 552,15.16.30-5; 553,5; 556,15.27; 563,30.33; 564,14.25.26; 565.10; 566,21; 570,20.24; 571,3.4; 573,7.15.17.21.28.32; 574,4.28.33; 575,15.29-32; 576,1.16.31-6; 577,14.21; 582,12-16.21-583.29; 584,5.11.16.26.27.30; 587,22.23; 588,14; 589,23; 595,17
antigraphon, transcription, manuscript, 522,24; 545,13; 728,10; 753,28; 769,17
antikrus, straight, outright, 521,8; 563,5
antimetastasis, successive displacement, exchange of position, 524,5.21.26; 531,32; 533,9; 543,8; 567,8; 572,34; 574,33; 581,22.27
antimethistanai, to exchange, 572,37; 575,1; 576,14; 577,17.19.20; 582,3
antimetrêsis, reciprocal measurement, 733,29
antimetrousthai, to be reciprocally measured, 768,23
antiperastasis, exchange, 596,10
antistrephein, to convert, 521,29; 580,16
antistrophê, conversion, 521,15.23; *to antistrophon*, the correlative, 710,11
antithesis, antithesis, 536,8; 554,23; 760,13.30
aoristos, indefinite, 536,25; 537,13.15.26.27.30; 538,1.9.12; 548,10; 571,34; 586,14; 593,31; 773,27.29
aparithmein, enumerate, 553,2.19; 555,31; 570,2.9
apathês, impassive, unaffected, 598,24
apeiria, unlimited number, 575,20
apeiros, unlimited, 519,14-16; 520,18.21.25.29; 521,6; 528,10.14.24.29.30.33 *passim*; *eis apeiron, ep' apeiron*, without limit, *ad infinitum*, 519,15; 534,11; 548,4; 551,14; 562,5.10.35; 564,17; 588,11; 593,30; 695,8.26; 705,11; 730,2, 739,33.34
aphairein, to abstract, 533,15; 538,22; 539,4; 551,30.32; 573,30; 574,4; 582,3; 767,12.13
aphairesis, abstraction, 537,13.16; 582,17
aphistanai, to avoid, shirk, stand away from, 523,14.27; 598,16; to emanate (away from oneself), 556,11
apithanos, implausible, 700,29
aplanês, fixed, fixed heavens, fixed sphere, fixed stars, 588,7; 589,20.22; 590,17; 592,3.5.17.19.28.31; 593,3.8.9.15.22; 594,19.22.26
apodeiktikos, probative, able to be demonstrated, demonstrative, 520,25; 521,8; 523,19; 696,1.12; 708,28; *apodeiktikôs*, demon-

stratively, 527,9; by way of demonstration, 564,34
apodeixis, demonstration, 522,22; 524,30; 533,2; 553,31; 555,20; 556,20; 561,32; 571,33; 754,4; 756,4.34
aporia, problem, puzzle, difficulty, 522,19.27; 528,27; 529,7; 534,9.14; 548,28; 549,6.12; 550,1.7; 221,13; 553,19; 562,7.12; 563,3.6.11.18.33; 564,2.18.20; 580,20 *passim*
apostasis, standing away (from *aphistanai*), distance, 756,24.28.31
apostêma, gap, 750,15
aretê, excellence, 552,15; 599,8
arithmein, to number, count, enumerate, 712,5.6; 713,19; 714,4.6.11.13.16.19.20.34.36.38; 715,1.2.7.8 *passim*; *arithmoumenos*, enumerated, 712,6; 714,4.6.11.13.16.19.20.34.38; 719,19.20; 720,35; 722,33; 728,28.30; 729,9.12.15.19.27; *arithmoun*, enumerator, 714,20; 715,8; 721,13; 729,13.19.23; 730,5.18.32; 731,29; 757,9.10; 759,24.26; 760,27; 763,2; 765,6.7.23.24; 768,32
arithmêtikos, enumerating, enumerable, 716,18; 759,3.4
arithmêtos, enumerated, enumerable, 713,21.33; 716.19; 719,19.20.24; 721,6.9.12; 722,26; 723,33
arithmos, number, 519,23; 535,30.33; 537,33; 540,13.18; 545,23; 585,14.17; 699,12; 702,30-32; 703,8.9.16.19.24 *passim*
arkhêgikos, originative, 527,22.32.35; *arkhêgikôtatos*, most originative, 522,22
arti, just, suddenly, 629,29.32; 752,9.11.14.17-23.28,29.33; 754,1 *passim*
asômatos, incorporeal, 521,14; 522,1; 527,25; 529,3.9.23.24; 530,5.6; 531,10; 532,3.5.20; 534,31; 561,2; 582,11; 593,36; 596,23; 600,19.22.27-9; 763,5
athroisis, conjunction, 560,8
aülos, immaterial, 538,1; 546,15; 597,23
auxêsis, growth, 534,22; 549,17; 567,15.17; 580,19; 591,20.30; 600,9; 711,32; 717,2; 720,25; 724,21; 737,23; 755,2.3; 760,25; 762,3.14.19; 764,25; 768,7.10
axiôma, assumption, postulate, presupposition, 565,9; 566,19; 569,30.32; 583,7.11; 585,4; 597,6
axioun, to assume, claim, postulate, 563,24; 565,21; 700,30; 701,15; 705,1; 708,35; 709,22; 725,4; *to axiôthenton*, assumption, 566,26; 571,16; 578,30.33

barus, heavy, 525,24; 546,28; 585,9
barutês, heaviness, 525,14; 538,13; 760,20
bathos, depth, 530,6; 542,1; 582,33; 587,18; 596,24
bia, force, *biāi*, by force, 566,11; 597,15.29.30.33

deiknuni, to show, prove, 519,9; 521,7.33; 522,9.11.29.34; 523,33; 525,3.11.21.22.32; 526,27.34 *passim*; *apodeiknuni*, to show, prove, 547,16; 548,9; 549,26; 573,32; 725,33; *endeiknuni*, to show, point out, 524,16; 540,16; 549,12; 552,27; 704,23; 705,9; 735,38; 761,1; *sunapodeiknuni*, to demonstrate, 525,1; 534,28; *hupodeiknuni*, to indicate, 529,8; 572,21; 583,30
deixis, proof, 524,21; 756,5; 757,1; 758,23
demiourgein, to construct, 704,21
demiourgikos, constructive, 704,23; artificial, 755,13
dêmiourgos, demiurge, 704,7.20; the god (meaning the demiurge), 707,24
diairein, to separate, 577,37; 726,29; 727,1-3; 748,9.11.21.23; *diaireisthai*, to be separate, discrete, 551,10; 559,30; 560,36; 566,1; 567,22; 569,25.26.30;

570,20.29.31 *passim*; *diairoumenos*, being divided, 591,27; 728,16; *diêirêmenos*, separate, discrete, 560,36; 569,16.18.25.26.30; 570,29.31; 571,3; 573,16; 576,16; 577,35.38; 588,18.19; 589,11; 591,15; 593,4; 598,19.21.27; 714,11; 727,19; 762,1

diairesis, division, 530,4.8; 551,12; 556,20; 562,7; 568,6; 569,24; 571,18.32.33; 697,4; 698,18; 700,23; 726,16; 727,1; 728,13.15; 729,11; 748,14; 749,4.18.22.26.28

diairetos, divisible, 695,8; 697,29; 698,26; 705,11; 711,12; 734,5

dianoêtikos, mental, 707,17

dianoia, thought, 707,14.18; 709,2.3.8; *kata tên dianoian*, mentally, 707,8; 709,9

diapherein, to differ, 524,22-4; 525,29; 538,1; 544,19; 550,4; 552,34; 556,5; 558,37; 559,22; 562,19; 576,25; 592,3; 697,2; 702,22; 712,28 *passim*

diaphora, difference, differentia, 524,9; 525,9.16.31.36; 526,4.6.23.29; 530,18; 531,25.30 *passim*

diarthroun, to articulate, 524,20; 583,18; 732,8; 746,12; 750,12; 754,2; 757,11; 770,15

diarthrôsis, articulation, 520,15; 753,32; 763,34

diastasis, extension, dimension, interval, 525,8; 531,9.10; 537,32; 538,3.4.9.13; 550.15; 554,16.18

diastatos, extended, dimensional, 573,11; 600,20-29; 697,16.20; 710,23

diastêma, interval, extension, dimension, 520,31; 529,27-32; 536,24; 537,5.10-31; 538,10; 539,8; 540,5; 567,17 *passim*

diathesis, state, 562,20

diïstanai, to set apart, stand apart, to be extended, to have dimension, 529,33; 536,1; 538,28; 542,6; 543,33.35; 548,33; 596,29.30; 600,19.22; 716,25; 756,25; *diestôs*, extended, 538,28; 596,29.30; 600,22

diorismos, distinction, 550,35; 555,33; 599,3; 713,5; 749,29.31; 766,24

diôrizein, to define, distinguish, separate, 520,20; 521,5; 526,23; 535,24; 538,3.8; 639,26; 559,31; 568,32; 571,8; 575,18; 583,18; 697,33; 698,1; 701,3; 709,12; 711,16; 712,31; 713,2.3.8; 716,27; 726,29.30; *diôrismenos*, separate, discrete, 526,24; 575,9.10.16; 576,4; 577,25; 710,29; 726,12; 729,36; 730,7; 756,1; 772,21

doxastikôs, based on opinions, 527,10

dunamis, power, capacity, 524,16; 525,1.3.9.20; 527,12,22; 533,17; 538,18; 542,5; 543,23; 578,32; 591,14; 704,1.15; 749,13; 757,2.6; 758,11; 759,8; *dunamei*, in potentiality, potentially, virtually, 520,29; 525,29; 535,14; 540,10; 541,6; 552,10; 567,33.34.38; 568,2.7; 577,39; 591,13.23.26.28; 598,36.38; 599,1-3,13-19.25; 748,21-4; 749,8-13; 757,24.27.30-2; 758,2.7.13-16; 759,8; 760,8

dunasthai, to be able, to be possible, 526,8; 540,16; 552,35; 573,12; 707,10; 730,16; 747,24; 755,15.16; 758,3; 759,11.27

dunatos, possible, 524,20; 525,14; 526,25; 529,3.10; 530,22; 532,8; 539,26; 543,1.16.18.21; 544,1.4 *passim*

duskolia, discomfort, uneasiness, 523,22; 566,28; 581,7.9; 596,2.5

duskolos, troublesome, 523,10.13; 565,5; 566,30; 581,13

êdê, already, 529,33; 533,11; 537,13; 552,4; 556,15; 558,26; 560,22; 565,9; 570,29; 571,11.16 *passim*

eidopoia, specific nature, 525,30

eidopoiein, to specify, 519,21; 524,8; 525,7.9

eidos, form, 519,5; 520,4; 523,4.5.9; 525,8; 525,8; 532,19.22.28.30.33; 533,1.3.7-9; 535,3.9.14.16

passim; species, 535,30.32; 536,3; 551,26-34; 552,1-3; 560,2 *passim*

einai, to be: *to ti esti*, definition, 520,26; 529,2.10; 531,13; 534,32; 535,7; 696,3; 700,13.16; *to einai*, being, essence, 520,32; 525,18.24.28; 527,14; 529,17-21; 531,14; 534,15; 542,12; 543,6 *passim*

ekei, there (see n. 56 to the Translation), 533,30; 546,11.13; 548,12; 549,18; 553,3.24; 587,17; 598,17; 705,20.32; 715,32; 758,21.31; 766,11; 772,3.10

ekhein, to have, possess, contain, surround, 520,8; 521,10; 522,27; 523,16; 524,16.35; 525,2.3.9.18.20.25.28; 526,2.5.14.17-29 *passim*; *ekhesthai*, to be a condition, 757,15; *ekhomenos*, be adjacent, contiguous, following, 580,26; 694,30; 695,12; 698,21

ekkhusis, diffusion, scattering, 538,1.13

ekstasis, departure, 740,22.25; extension, 536,25

ekstatikon, variation, departure, 711,14.18; 753,30; 754,4.17

endidonai, to implant in, 734,20; 742,8; 766,17

endoxos, received opinion, 521,7; 528,9; 696,1

energeia, actuality, activity, 598,32; 704,25; 735,13.33.38; 736,18; 737,3.24.28.29; 742,3; 754,27; 758,11; 763,6; *energeiāi*, in actuality, actually, 531,29; 541,6; 567,23.32.38; 568,3; 570,7; 575,19; 591,13.23.26; 599,15-17.25; 735,34; 748,24; 749,9.10.12; 757,25.27; 758,8.12.13.17; 760.7; *kat' energeian*, in actuality, actually, 575,21; 591,15.16; 739,28; 757,6.22.34.35; 758,5

energein, to activate, be active, 708,34.35; 709,3

energos, active, 704,22

enistanai, to be present, 698,11.12; *enestôs*, present, 697,8; 724,29; 731,7; 732,3.4; 750,15; 751,8.21.31; 752,2.7.10.12; 756,25; 764,4; 771,18; 772,15.31; 773,4

enmethodôs, methodically, 522,31; 529,12

ennoein, to conceive, have a concept, 529,12; 703,31; 707,9.13.16; 709,17.18.21; 713,10; 716,6; 749,29

ennoia, thought, concept, conception, view, idea, 524,20; 527,5; 528,35; 531,19; 535,10; 537,26; 538,2; 539,12; 544,15.17; 696,13; 700,26; 705,8; 707,11.16; 708,4; 709,1.6.19.24; 710,5.8.10; awareness, 709,6.19; 710,5.8.10

enstasis, objection, 525,21; 578,11; 597,8; 718,15.18; 743,11; 750,26; 757,20; 770,15

entelekheia, actuality, realisation, 540,10; 712,21; 735,35

entha, ***enthade***, here (see n. 56 to the Translation), 531,20; 548,12; 599,15; 755,16; 758,20

epagein, to adduce, bring forward, to lead on, 527,29; 528,30; 529,4; 532,29; 535,10.11; 536,27; 538,5; 541,28; 546,21; 547,24; 552,16.19.25; 555,29; 558,27; 564,33; 578,22; 580,1.20 *passim*

epagôgê, induction, 543,8; 555,32; 556,17

epaktikos, inductive, 543,9

epharmottein, to apply to, fit, coincide, 531,28; 550,35; 557,2.20; 561,13; 569,31; 593,22; 596,22; 763,17.20

ephexês, adjoining, 519,18; 521,6; 522,29; 524,3; 546,31; 554,30; 566,17; 567,4; 570,28.32.36; 571,18; 572,31; 582,24; 594,20; 596,11; 597,10.14.25-27.33 *passim*

ephistanai, to pay attention to, consider, 588,4; 709,14.22; 728,29; 730,12

epikheirêma, argument, 521,7; 522,10; 524,11.13.17.19.36; 526,22.26.27.34; 527,4.9; 528,29; 529,1.4.18 *passim*

epinoein, to conceive, think, 591,28; 704,4; 707,11; 713,16; 748,10; 749,10
epinoia, concept, thought, 526,8.20; 574,4; 709,23; 715,6; 717,6; 766,7.23.27; 767,12; *epinoiāi*, conceptually, 717,20; 749,8.12.26; 766,12.28.32; 767,14
epipedos, plane, 538,11.16.19.25.28; 539,1.3; 587,17-26.33; 596,22; 699,15-17
epiphaneia, surface, 530,29; 531,17.22.31; 536,5; 538,26.29; 543,27; 548,27; 554,6.12.17.19.28.32; 555,2.14-17; 562,23.25.27 *passim*
epistasis, attention, consideration, 710,5; 715,11; 724,27
epistêmê, learning, knowledge, 546,4-9; 554,7; 564,24; 567,28.35; 717,10
epistêmôn, one who knows, learned one, 554,6; 555,5; 707,1; 754,8
êremein, to be stationary, to rest, 519,21; 533,13.21.22; 567,19; 578,25-8; 580,11; 581,27.33; 582,5; 727,26; 742,31.32; 743,18.21.24.29.30; 744,9.23.24; 745,4.6.8.11-15.18-22.26.28.32; 746,24.26.31.33; 757,28.31-3; 758,1.4.8
êremia, rest, 702,23; 742,22.23.27.29.33; 743,3.4.7-10.17.22.31.35; 744,2.6.11.13.16.18.20.22.25-8.32.35; 745,1.7.14.24.32; 746,3.5.1.24-32; 750.25.27.30.31; 757,34.35; 758.6
erôtan, to ask, 564,5; 751,3; to use an argument, 521,19
erôtêsis, question, 564,5; supposition, 521,12
euapoblêtos, easily lost, 562,20
euthetismos, ordering, 739,36
euthus, straight, 519,12; 522,32; 528,35; 573,10; 581,8; 709,21; 710,6; 713,10; 752,17; 755,33; 756,22; 770,28.30; 772,12; 773,26
exaiphnês, suddenly, 753,1.3.4.7.9.21.27.28; 754,1
exallassein, to change, 576,17
exarithmein, enumerate, 718,15
exêgêtês, expositor, commentator, 543,30; 545,7; 553,2; 559,17; 590,23; 707,33; 755,1
exetasis, examination, 560,12; 762,29
exetazein, to examine, 581,26; 762,27
exisazein, to make equal, to be equivalent (intrans.), 521,20.23; 536,13
exôterikos, nontechnical, *dia tôn exôterikôn*, nontechnically, 695,34

genesis, becoming, generation, 525,12; 528,16.23; 534,22.26; 535,32; 539,11.21; 540,1; 544,26; 546,6; 566,15; 589,31; 597,20.22; 599,4.6.21; 600,36 *passim*
genêtos, created, generated, 539,18; 541,34; 704,3; 744,28; 746,34; 768,6.8
genos, genus, kind, 528,28; 529,23.22; 530,5; 539,20; 542,6; 546,4; 551,26.27.30-4; 552,1-4; 556,6-9; 560,3
ginesthai, to become, 520,30; 521,24; 523,13.15; 525,27; 530,30; 531,20; 532,2; 533,31; 534,23-5; 537,2 *passim*; *ginomenos*, coming to be, *to ginomenon*, thing that comes to be, 523,20; 534,3.22; 539,18; 549,14; 550,15; 557,11; 564,25; 570,12; 578,31; 579,9; 581,19.30; 598,28.36.37; 699,34; 705,18; 713,16; 715,33; 720,23; 723,8.16; 724,1.3.30; 733,3.7.19; 739,24; 741,7; 742,7-12; 747,28; 748,6.30.34; 753,5.27; 754,5.26.31; 755,11; 765,8; 768,9; *en tôi ginesthai*, in becoming, 520,30; 696,31; 701,11; 705,9; 713,33; 720,5.9; 721,18; 724,16.32.34; 725,16; 731,23; 735,18.24.25.29; 736,8; 737,8.16-18.29.33; 739,20.24.27; 744,16; 761,4; 763,16
gnôrimos, knowable, recognisable, familiar, well-known (see n. 256

to the Translation), 717,14; 724,10.13; 725,32; 768,5.10; 768,26.29-31.34; 768,36; 769,3

graphê, text, passage, reading, manuscript, 545,12; 570,30; 589,34; 697,23; 728,11; 729,9; 734,8.12; 748,17; 758,9

graphein, to write, 522,26; 526,9; 545,17; 558,34; 563,23; 569,25; 589,33; 595,5; 710,35; 723,37; 728,1; 739,22; 766,20

hama, simultaneous, at once, 531,1.2; 534,28; 539,32; 544,26; 548,9.37; 549,2; 558,28; 559,2.15.17.33; 561,21.31 *passim*

haphê, touch, 599,23.31

haptesthai, to touch, to be touched, 568,11.13.14; 572,17; 573,11.13.30; 574,3; 588,37; 589,26; 590,31; 591,27; 592,23.26; 597,28-33; 599,28; *haptomenos*, touching, being touched, 548,27; 568,15.18.20.22.31; 569,18.34; 570,5; 573,29.30.33; 574,2; 594,12.24; 597,14.25; 598,6.10; 599,35

haptos, tangible, touchable, graspable, 539,21; 541,27.34

hêgeisthai, to think, 521,14; 704,17; 707,10.14.22; also, to lead, hence: *hêgoumenon*, antecedent, 521,21.24; 522,13

hepesthai, to follow upon, *hepomenon*, consequent, 521,21.24; 522,14; following, 593,17

hexis, condition, disposition, 530,22; 543,11.16-20.23; 544,4; 552,25; 556,33; 562,18.19.26; 732,32; 742,24; 746,33; 757,13.19.30; 762,16.19.22; 765,12-14; 766,4.21; 767,7

histanai, to be stationary, 745,4; to come to a halt, 562,10; 564,13; 727,5

holikos, universal, general, *holikôteros*, more general, about place: more voluminous (a larger place than one's private place), 536,14.33

holos, whole, entire, *to holon*, the whole, 530,28; 531,15; 538,21.30; 543,16; 547,12; 551,25.30-34; 552,1.3.5 *passim*; *en holôi*, *en tôi holôi*, in a whole, in the whole, 536,31; 551,18.19.29; 552,32; 556,2; 559,25-33; 560,6.8.16.19.24-9.36.37; 561,1.4 *passim*; *holôs*, entirely, in general, generally, at all, any, 521,22.34.35; 523,16; 525,34; 526,25; 529,5.8; 530,3.4.27 *passim*

homoiomerês, having similar parts, *ta homoiomerê*, things having similar parts, homeomeries, 532,12; 551,33; 577,39

homônumia, ambiguity, 701,2

homônumos, ambiguous, 553,11; 539,17; 553,12; 559,30

horismos, definition, 519,12; 520,5; 522,34; 529,23; 554,14; 556,25; 557,2.20; 561,13; 565,6.7; 566,27.29; 567,4; 568,20; 581,8; 582,29; 584,17; 596,1.3.6.26.32; 597,7; 714,5; 717,22; 718,16; 719,7.18; 734,34; 743,34

hormê, impulse, 566,16.18

hugiainein, to be healthy, 552,14

hugieia, health, 552,7

hugiês, valid, 521,35; sound, 548,9; 759,22

hulê, matter, 519,5; 520,4; 523,6.7.9.21; 532,19.22.28.30.33; 533,1-6 *passim*

hulikos, material, 536,24.25.29; 537,24; 538,4.9; 550.15; 755,14; *hulikôteros*, more material, 599,5.24; *hulikôtatos*, most material, 531,6

huparxein, to belong to, apply to, 556,25; 559,22

huparxis, existence, being, 527,29; 542,27; 566,28; 585,13.16.20; 596,3; 695,15.16.21.25.27; 697,31; 712,25; 724,13; 735,33; 736,9; 741,20; 744,19; 747,18

huperbibazein, to transpose, 521,18

huperouranios, supercelestial, 522,8; 541,10.11.25; 546,1

huphistanai, to exist, to be real,

525,20; 527,3.8; 535,11; 697,8; 704,6; 707,6.7; 724,15.17.34; 730,10; 735,28; 739,19.21.27; 760,30; 762,31; 765,26; 766,24.26

hupodeiknunai, to indicate, 529,8; 565,9.11; 567,13; 572,21; 582,6; 583,30; 584,16; 702,34

hupodokhê, receptacle, 520,2; 534,27; 539,11; 541,26.33; 545,27; 579,14.16

hupokeisthai, to underlie: *hupokeimenos*, underlying, 755,1; *to hupokeimenon*, substrate, 535,6; 537,29; 542,18; 544,27; 551,24; 552,8.18.20.21.28.29; 553,4.9 *passim*

hupomnêmatistos, commentator, 533,32

hupostasis, existence, 520,34; 528,34; 542,3; 545,1; 562,14; 565,20; 574,11; 695,4; 700,7; 716,12; 717,16; 728,8; 736,8; 739,23; 749,28; 750,31; 766,13.27; 769,11

hupothesis, hypothesis, 530,11; 531,8; 538,27; 541,20; 548,6; 559,7.14; 561,17.25; 572,11; 573,22; 574,15.16.20; 577,22.25; 578,3.24.33; 581,26; 588,13; 701,15; 704,29; 705,3; 721,7

hupotithesthai, to propose, assume, hypothesise, 520,21; 524,18; 526,34; 573,23.25.27.28.29; 574,9; 576,20; 582,8; 701,31; 704,25; 705,22; 725,27; *hupotithesthai to zêtoumenon* (lit., to assume the thing sought), to beg the question, 573,22

husteros, later, after, 530,10; 571,10.24; 698,8.17.22; 699,28.33; 704,20; 708,16.25; 709,15; 710,20; 712,4.9.12-32; 713,5.7.12-21.32 *passim*

idios, proper (*to idion*, thing proper to), private, peculiar, 526,22.25; 533,21; 535,15.22.23; 536,7.10.13.20.32; 542,7; 544,5.10 *passim*

kata, according to, with respect to: *kata meros*, bit by bit, 520,30; 554,24; 555,3.13.25; 557,29.32; 558,2-6.15.16; 560,35; 592,33.34; *kata skhesin*, relatively, relative, 525,16; 526,7; *kata skhesin pros hêmas*, relative to us, 525,17.19.28.33.35; 526,1; *kath' hauto*, as such, in itself, essential, 519,8; 520,11.31; 523,1.2.10.24; 526,13.20.21; 527,14; 535,1.5.26-35; 536,1-17 *passim*; *kath' heauto*, in itself, as such, 532,23.24; 533,26.29; 568,4.11; 579,9; 711,10; 766,13; 767,13; *kat' ousian*, essentially, in essence, 557,15; 566,6; 595,17; 742,2.4.12; 761,18; 705,18; 726,2; *kata poion*, qualitative, of quality, 595,18; 761,18; *kata poson*, quantitative, of quantity, 595,18; 706,14; 761,18; *kata sumbebêkos*, contingently, 519,9; 526,12; 535,28; 536,6.14; 548,20.23.24.33; 549,3; 554,24; 557,24.29-33; 558,1-38; 559,2.13.25 *passim*

katalêpsis, grasping, understanding, 523,11; attainment, taking possession, 533,20

katametrein, to measure out, 697,11.13; 736,10.12.23-27.30.33

katanoêsis, understanding, 523,13

katatattesthai, to be ordered, numbered, 721,19; 765,18; to be instantiated, 720,36; 721,25

katêgorikôs, categorically, 547,28; 697,3

keisthai, to be situated, established, lie in, 540,24; 541,8; 552,12; 587,22

kenembatein, to be vacuous, 582,18

kenos, empty, *to kenon*, void, 520,12.14.31; 524,13.14; 526,13.14.32; 527,1.5; 528,2.3.8 *passim*

kenoun, to be emptied, 572,37; 573,8

kentron, centre, 531,3; 585,12,34; 590,9.16; 709,17

khaos, chaos, 523,18; 527,18.19.24;

528,13.22.25; 529,16; 534,1
khôra, space, location, place, 521,27; 523,5.7; 524,28; 527,26.29.30; 528,13; 531,14-16; 539,6.10.20.23; 540,1.5.21.33; 541,26.35; 542,5; 545,25; 546,16; 567,20; 573,18; 578,21; 581,18; 582,12.14; 583,32
khôrein, to spread, *khôrein dia*, to permeate, 530,9.13.14.17-25; 530,30; 531,3-6; 538,30; 540,33; 561,4; 562,6; 575,7; 577,33; 578,4.6; 582,8; 590,9; 590,13.19
khôrêtikos, accommodating, receptive, holding, 527,20; 543,32; 583,5; 588,15; 707,10; 708,32.37; 726,1; 748,26; 760,6; 765,26
khôrismos, separation, 544,24
khôristos, separate, 522,2; 543,13; 544,6.14.20.25.28; 545,27; 565,18.19.23.24; 567,10.36; 572,37; 578,9; 591,15; 767,9.21; 769,10.11
khôrizesthai, to be separated, 543,3-7.15-17; 544,3.26.28; 565,26; 567,35; 712,27; 749,10
khronikos, temporal, in time, 716,1.8; 719,12.29; 738,10; 750,11; 751,35; 752,11; 753,6.32
khronizesthai, to become temporal, to be timed, 716,11.12; 720,10; 742,15; 744,29
khronos, time, 519,20-5; 520,1.6.14; 522,23; 528,15; 553,6.8; 694,31; 695,1-6.10-16.20.26.31-3
khusis, diffusion, 536,26; 537,26
kinein, to move, change, *kineisthai*, to be moved, changed, (intr.) to move, change, 519,21; 520,10-12; 522,21; 525,6.26; 532,32; 533,12; 546,26.27; 547,9.11.14.21.27-32; 548,10.15.17.18.30; 550,34.36; 552,10; 558,2.9-14; 566,12; 567,6.7.9.21.22.26.29-36; 568,1.3.6.8-12.17-22.28.32 *passim* (Most occurrences of forms of the verb 'to move' are translations of forms of either *kinein/kineisthai* or *pheresthai*. When not translated as forms of 'to move', forms of *kinein* and *kineisthai* are translated as forms of 'to change'. Occasionally, forms of the verbs 'to move' and 'to change' translate other words of change and motion. See n. 1 to the Translation for a list of these.)
kinêma, movement, 707,22; 713,4; 725,25; 728,17; 729,7; 763,8
kinêsis, motion (complete list), 525,11; 534,2; 547,1.3.4.17-19; 566,8.16; 570,17; 574,30; 589,23; 590,25; 591,18; 593,21; 700,17; 701,17.24.32; 702,2.10-14.21-23.28; 703,21.22; 704,12.26.27.31; 705,2.6.12.20.21.23.26.28; 706,34; 708,26; 712,12.13; 715,18.22.24.28.36; 716,5; 719,2; 721,9; 722,29.32; 723,3; 725,25; 726,18.24.25; 736,20; 743,16; 754,19.22.23; 756,2.5.21; 757,3; 758,24; 760,16.21; 764,31; 768,35; 769,5.20
kinêsis, change (incomplete list), 519,10-14.22-6; 520,1.3.5.11.14.15.18.20.23; 525,1; 533,16; 542,7; 567,12-14; 568,23; 576,11; 580,8; 588,21.25.26; 595,17.19.21.26.35 *passim* (Most occurrences of the term 'change' translate *kinêsis*. See n. 1 to the Translation for a list of other words of change and motion.)
kinêsis kata phoran, locomotion, change of place, 591,17.18; 593,20.21
kinêsis kata topon, local motion, 522,12.15.18; 567,12-14; 580,8; 588,21; 595,19.22.26; 705,21.22.25; 711,31.35; 712,2; 715,26; 716,36 *passim*
kinêtikos, cause of change, 533,17; 755,6
kinêtos, moveable, changeable, 580,12; 584,26; 594,13.25; 595,29; 704,4; 712,21; 735,35; 757,17.18.26.27; 758,11.14.18.22.23
kinêtos kata phoran, that which

changes by locomotion, 579,36; 580,7.9.10.17.19; 585,21; 588,21; 596,37; 597,1

kinêtos kata topon, that which changes place, 580,33; 591,29

koilos, concave, 570,24; 572,36; 573,30.32; 580,5; 585,16.20; 586,25; 589,20; 590,28.32; 594,14.27; 751,10.27

koilotês, concavity, a hollow, 586,10

koinônein, to share with, to be associated with, 544,18; 721,29

koinônia, affinity, having things in common, 707,3; 738,28

koinos, common, 519,4; 520,15; 520,8.9; 522,11.14.15.19.25; 528,34; 529,7.15 *passim*; *koinê ennoia*, common conception, 528,24; 550,27; 565,8; 566,24,26; 567,3 *passim*; *koinê hupolêpsis*, common sense, common understanding, 521,8.9; 545,5

kosmos, universe, 535,19; 569,5; 574,12; 576,20; 582,8; 588,8.27; 591,11; 593,8.32.33.36.67; 594,7.9.18.20.34; 701,30; 704,14.24

kuklophorêtikos, revolving, rotating, 585,14.33; 589,9; 590,20; 594,10.35

kuklophoria, revolution, circular motion, cyclic motion, 595,17.26; 741,22; 751,1; 760,16.19.23.24; 768,10.23.25.28.30; 769,2.5.27.31.33.35

kuriôs, in the basic sense, strictly, properly, primarily, absolutely, 522,1; 535,17.23.27; 536,32; 540,24.25; 552,28; 554,24.25; 555,17 *passim*; *kuriôteros*, more crucial, more important, 527,7; 553,13.15; 560,29; 738,29; *kuriôtatos*, most basic, 522,21.24; 551,2; 552,29.31; 554,23; 555,19; 563,29; 564,31; 571,15; 584,31; 750,33

lexis, text, passage, words, 521,18; 526,16; 563,9; 569,27; 576,29; 586,28.31; 702,2; 710,32; 714,33.37; 722,11; 739,22; 758,25

logikos, rational, 536,6; 556,20; 771,8; 772,7

logistikos, rational, *to logistikon*, the rational part, 554,6; 555,1.5

logos, account, argument, reason, reasoning, 519,17; 520,14.16.27; 521,6.28.31; 522,3.4.10.13.29; 523,15; 524,2.4 *passim*

manôsis, rarefaction, 588,2; 589,2

megas, great, 524,16; 542,12; 545,24; 581,3.9; *meizôn*, greater, 524,33; 530,27; 531,7; 549,13; 550,13.18.23; 559,35; 565,16; 567,16.17; 576,18.19; 590,18; 596,12; 733,21; 740,3.4; 745,20; 756,22; 762,20; *megistos*, greatest, 530,20; 576,19; 738,34

megethos, magnitude, 532,2.6.14.15; 536,24-30; 537,10.12.18-32; 538,4.5.9-13.23.24 *passim*

mêkos, length, 530,6; 554,19; 562,28; 587,18; 596,24; 600,26; 711,2-7; 726,33; 727,2; 748,10.33; 749,18; 770,8

mellein, to come, to be about to, *mellôn*, coming, *to mellon*, *to mellein*, the future, 697,7.31.32; 698,2; 699,1-4; 700,7; 701,10; 713,6.11; 720,10; 713,6.11; 720,10; 731,31; 732,5; 736,2; 746,18; 747,5.9.14.15; 748,5.8; 750,2.17; 751,9.17-23.32; 752,16.27.29; 756,26.29; 759,8; 764,6; 771,20

menein, to remain, rest, 533,10; 534,25; 545,23; 557,14; 561,6; 566,10; 573,12.16.21 *passim*

merismos, division, 561,10

meros, part, 520,30; 524,11.22.23.31; 525,8; 526,10; 531,27; 543,24; 548,20; 549,14; 551,18-25.29-34; 552,1-3.21.23.32 *passim*

mesos, intermediate, in between, 525,5.6; 536,4; 559,28.37; 560,25; 563,10.12; 584,29.34; 585,1-6.9-12.30; 586,3-8.14-16.26-35; 587,6-9.12; 599,34; 600,3.4; 712,33; 716,8; 727,25; 735,10;

742,17.18; 751,31
metabainein, to move on, change position, exchange place, 529,2; 535,7; 567,25,30; 568,15; 576,26; 577,5; 578,8; 580,14.23; 582,6; 585,18; 588,33.34.38; 590,34; 591,2.36; 736,38
metaballein, to change, exchange, alter, transform, 523,32; 524,29; 533,12; 548,9.15; 546,15.23; 550,30.32; 567,2.37; 571,35.36; 572,1; 574,25; 575,28; 578,3; 582,7; 585,14; 588,29; 589,3; 590,1.2; 591,6; 597,13; 598,9; 599,17.20.26; 600,5.7.8; 705,13-20.29.33.36; 707,18; 709,2.4.8.12; 739,28; 740,24; 742,16; 745,3; 754,21; 756,15
metabasis, change of position, 591,8.19; 724,17; 744,7.14
metabolê, change, transformation (see n. 163 to the Translation), 549,8.27; 550,3.11.13.17; 597,17.20; 599,5.20; 704,31; 705,3.12.15.16.28; 706,1.28-34; 708,22.26.36; 709,3; 731,19.20; 737,23; 740,23.25; 741,21; 753,26; 754,4.17.19.22.23.34; 755,21.33; 756,2.5; 757,3; 760,16.20.24.25; 761,19.24.25; 762,20
metalambanein, to partake of, participate, 528,13; 536,9; 540,19.31.32.34; 542,11; 544,27; 563,10; 593,24; 712,10; 715,20; 758,32
metalêpsis, participation, 541,5
metalêptikos, that which partakes, 541,2.13; 542,10
metapheresthai, to transfer, 573,4; 575,32; 711,30; 723,13
metaphora, transfer, 576,36; *kata metaphoran*, metaphorically, 540,26; 763,5
metaphorêtos, moveable, transportable, 543,33; 552,16; 582,26.34; 583,1.6; 584,26
metaphorikos, metaphorical, 540,22.23; 540,28; 553,16
metastasis, shift, change (of place), 550,30; 567,11.18
methektikos, that which participates, 545,20.22.23
methektos, participated, 729,20; applied, 763,3; 765,29; 767,10.17
methexis, participation, 761,21; 766,23; 772,4
methistasthai, to change, vary, alter, move, shift, exchange (with respect to place), 533,9; 541,18.19; 548,11; 567,16.20; 572,17.35; 574,31; 575,15.29; 576,2.5.12; 577,7.14.22; 740,2
metrein, to measure, 519,23; 537,23; 697,17.21.22.29; 714,1.6.31.35; 716,28.33.35 *passim*
metrêtikos, measuring, 742,24; 743,7; 747,8
metrêtos, measurable, 697,20; 736,13
metron, measure, 520,3; 528,16; 536,27; 703,22.25.29; 714,6; 716,9.14; 722,1; 733,10.20.23.31; 734,31 *passim*
mikros, small, 542,12; 545,24; 579,6; 710,32; 715,10; 722,11; *mikrotatos*, smallest, 530,19
mikrotês, smallness, shortness, 753,2.10
monadikos, abstract (made up of ideal units, see n. 178 to the Translation) 714,7.35; 715,1; 717,31; 718,6; 738,18; 762,2; 763,2; 764,8.20; 767,19; 772,33
monas, unit, 708,35; 715,2.3.6; 722,2; 725,12.22.30; 726,5.6.12; 728,22; 730,8.14.15.19.21.30 *passim*
monôs, simply, 528,34; only, 568,9; *monôs kai telôs*, pre-eminently, 567,14
morion, portion, 543,11.14-22; 544,3; 551,20; 553,18; 554,1.2; 555.6; 567,29; 568,28; 574,32; 575,1-23
morphê, shape, form, 536,23.26; 538,16; 542,18; 548,7; 552,26.29; 579,9.10; 586,2; 715,4

noein, to think, conceive, 523,25; 526,18; 538,13; 574,34; 579,31;

703,31; 707,7; 708,29.32.33.36; 709,10.22; 710,4; 712,30; 713,1; 717,28; 718,7; 723,37; 724,3; 738,29; 748,33; 749,4.5; 750,30

noeros, intellective, 738,4; 755,17

noêsis, thought, 539,17; 707,13; 708,31.32

nous, intelligence, 522,8; 759,4.16; 760,17

nun, now, 521,2; 524,6.27; 529,18; 530,21; 531,22.33; 532,5.36; 535,19; 536,11; 537,8; 540,11 *passim*; *to nun*, the now, the present moment, 522,6; 527,10; 540,17; 697,8.9.11.25.28.33.35; 698,2.5-7.18.21.27-32; 699,2.4.8.11.18.21.22 *passim*

oikeios, proper, specific, characteristic, related, relevant, 519,22; 523,27; 524,1; 526,14.15; 527,1.3.12.23; 533,20; 535,16; 536,22; 541,4; 547,32 *passim*; *oikeioteros*, more proper, relevant, apposite, 599,21; 748,19; 769,2; *oikeiotatos*, most nearly related, 597,21; of (most) immediate concern, 694,32

oikeiotês, relationship, 598,23.24

onkos, bulk, 530,27; 570,4

ouranios, heavenly, celestial, *ta ourania*, the heavens, 522,20; 531,7; 593,23; 703,21; 742,4.14

ouranos, the heavens, 521,28; 530,24; 531,3.4.8.9; 535,19; 539,24; 540,2.15; 545,19.31; 546,10; 564,6.10; 568,27; 569,5.7.8 *passim*

ousia, essence, substance, being, 520,26; 521,1; 526,8.11; 535,29.31.33; 536,2; 539,29; 546,3; 549,5; 551,5; 552,21.25 *passim*

ousiôdês, essential, 543,22; 585,23; 723,17; 737,28

palai, long ago, 695,18; 707,25; 752,7.31.32; 754,1

palaios, old, ancient, *hoi palaioi*, the ancients, 521,10; 522,4.5.9; 562,13; 565,32.33; 587,19; 699,29

paradeigma, paradigm, 526,22; 539,2; 552,18; 555,19.25.27.29; 557,4; 560,37; 565,11; 570,35; 571,3.5.6; 583,30.31; 703,32; 717,27; 739,16; 740,5; 747,19; 749,23; 752,15

paradeigmatikos, paradigmatic, 717,30

parakolouthein, to track, to parallel, to follow, 519,4; 533,14; 552,24; 695,5; 707,28

parakolouthêsis, (mental) tracking (see n. 169 to the Translation), 707,18

paralogismos, paralogism, 537,2

paratasis, extension, duration, 709,14; 716,8.22,26,33; 717,5; 719,1.15; 720,1; 721,20; 723,32; 731,13; 734,21; 735,27.33; 736,17.20; 737,32; 738,1.3; 742,6; 744,18.21.25; 746,27; 750,32; 753,6; 755,26; 767,24

parekbasis, digression, 546,16

parekteinesthai, to extend together, 729,27

parelêluthos, past, 696,25.34; 697,7; 698,1; 701,9; 713,6.11; 720,9; 724,29; 731,30.14; 732,5; 733,5; 747,6.13.15; 748,5.7.14; 750,2.16; 751,9.16-22.21; 752,11.14.19.29.32; 756,25; 764,6

pas, all, total, whole, *to pan*, everything, the totality, the universe, 528,15.21; 534,19; 564,6.7.9.12; 568,14.15; 580,30; 583,35; 585,5; 588,12.14.28.31; 590,1.5.7.9.12-16 *passim*

paskhein, to be affected, be acted on, 561,21; 582,32; 598,4-6.10; 695,13; 723,23; 740,13.17.27.28; 741,14.16.19; 754,30

pathêma, affection, change, 542,5

pathêtikos, affective, 562,21; 759,4; *pathêtikôs*, in the passive (voice), 697,17; *to pathêtikos*, the passive (voice), 697,20

pathos, attribute, affection, 538,23; 539,4; 543,19.23; 552,25; 558,10; 562,18.19.26; 738,14.15.20; 757,14.19.28.30; 765,12.13.15; 766,4.21; 767,7; experience,

753,13
perainein, to bound, 528,20; 696,1; 699,21; *peperasmenos*, bounded, finite, 519,14; 554,20.21; 562,28; 588,10; 593,29; 695,9; 699,13.18.20.22; 736,28; 739,32; 740,10.11
peras, limit, 520,11; 528,5; 529,25; 531,16.22; 535,16; 536,19-23.37; 537,14.15 *passim*
periekhein, to contain, 535,22.23; 537,1; 538,11; 544,10.12.13.29; 551,28; 553,24; 560,17.20-3.28 *passim*; *to periekhon*, container, that which contains, 523,5; 528,5; 535,13.15.17; 536,17-19.35; 537,3; 538,25; 544,20; 548,26; 561,14.16.18.19.20; 562,30.31 *passim*; *to periekhomenon*, that which is contained, 548,27; 560,24.29; 561,14.16.21; 569,1.16.29.34; 571,14 *passim*
periektikos, containing, *to periektikos*, containment, 523,3; 535,23.24; 536,20; 537,16; 541,3; 544,18.19; 571,14; 594,3; 738,27
periphora, rotation, revolution, cyclic motion, 580,13; 589,6.7; 591,19; 593,20; 595,9; 700,17.27; 701,17-31; 702,8; 703,1.6.10; 704,34; 705,4.7; 768,14.19.35.36
peristasis, circumstance, 553,10
phainesthai, to appear, seem, 525,16; 540,24; 541,30.33; 542,2; 563,18; 566,27; 574,15; 582,16; 697,33; 700,15; 709,28; 710,34; 711,3.5; 717,8; 723,9; 728,11; 732,27; 741,11; 752,28; 762,26; 765,33; *phainomenos*, appearing, apparent, 521,15; 540,4; 545,21.25; 564,19; 704,19; 755,20
phantasia, imagination, 574,5; 582,17.19; idea, 581,30; appearance, 707,34
phantasma, image, 539,28
phantasta, things imagined, 526,31
phantazesthai, to imagine, fancy, 573,9, 582,4
pheresthai, to move (locally), 525,5.24; 539,28; 542,8; 544,25; 545,13; 546,22.27-31; 547,9.14.15.21.23.27; 556,5 *passim*; *pheromenos*, moving, 546,30.31; 580,14; 585,7-9; 595,24; 597,24; 721,28; 722,29.31; 723,2.10.21-7.37; 724,14-25.30.35; 725,24.28.31.34; 726,1-4.7.12.17-31; 727,7; 728,10; *pephorêmenos*, set in motion, 539,18
phônê, expression, 553,12; sound, 742,25
phora, motion, locomotion, local motion, 519,22; 521,4; 522,23; 524,8.9.12.34; 525,15; 533,20; 567,14; 579,36; 580,7-11.17.19 *passim*
phthartos, perishable, 740,20; 746,34; 755,5.13
phthora, passing away, perishing, 522,16; 539,20; 544,24; 549,10.16.30; 550,7.19.27; 599,4.6.20; 600,36; 706,29; 711,33; 737,29 *passim*
phusis, nature, 519,9.12; 520,6.26; 521,14; 522,23; 523,3; 524,7.35; 525,1.2.12.17.22.23.27.31.32.35.36 *passim*
phusikos, natural, 519,3.4.10-15.19.20; 520,1.2.4.7.9.16.17; 521,7; 522,4 *passim*
pistousthai, to prove, 529,19; 554,5; 585,27; 707,30; 708,11; 719,25; 727,29
pithanos, persuasive, 700,12; *to pithanon*, persuasiveness, 578,23; *pithanôteros*, more persuasive, 521,19; 529,11; 572,11; 578,19; 581,25
planan, to wander: *to planômenon*, planet, planetary sphere, planetary area, 589,30; 592,8; 594,29
planê, a wandering (of the planets), 703,17.25
platos, breadth, 530,6; 554,19; 562,28; 587,18; 596,24; 600,26
platus, broad, 772,29; *en platei*, in the broad sense, broadly, 752,1.22.23.25; 772.29; in the colloquial sense, 747,34; *platuteron*, at greater length, 762,11

ploutein, to be wealthy, 552,14
poion, quality, 595,18; 706,2.19.21; 761,18; *poios*, what (inter. pr.), 538,14
pollakhôs, in many ways, *legesthai pollakhôs*, to be said in many ways, to have many senses, to be equivocal, to be ambiguous, 551,11; 553,11; 562,15; 563,24; 595,15; 696,13
pollakhou, in many places, frequently, 583,9
poros, pore, channel, 530,21
poson, quantity, 533,24; 536,2; 537,23-5.33; 538,29.31-3.35; 568,23; 595,18; 697,13.22; 698,15; 705,18; 706,2.15.22; 711,12.15-17.22.26; 714,35; 716,19; 729,36; 761,18; 764,19
pote, sometime, 522,27; 551,5; 565,3.19; 568,3.8; 574,12; 582,35; 600,12.15; 704,19; 745,5.28.32; 747,2.25; 749,12; 750,20; 760,4; *pote men ..., pote de ...*, sometimes ... sometimes (drawing a contrast), 535,34; 536,1; 571,7.8.29; 580,11; 583,19.20; 584,14.15; 586,15; 594,36; 723,14.15.27.28; 727,5.6; 745.9; 747,27.28; 748,9; *ho pote ên*, *ho pote on* (see n. 177 to the Translation), substrate, substratum, 712,25; 721,27; 722,13.19.20; 726,20; *ti pote*, what, 522,27; 551,5; 565,3; *to pote khronos*, sometime (the term or concept), 750,8.13; also *to pote*, 751,35
pragmateia, investigation, subject-matter, work, treatise, 519,3; 520,7; 595,25; 599,4; 706,29; 748,4; 750,25; 755,8; 769,5
proagein, to present, 528,25; to adduce, 545,31
proaxiousthai, to presume, presuppose, 585,4; 696,28
problêma, problem, question, 520,14.20.23.24.28; 521,5; 523,14.22.27; 524,1; 529,2.13; 535,4.8; 549,11.24; 564,5; 566,29; 696,3.12
proêgeisthai, to precede, be pre-eminent, 519,25; 593,16; 696,11; 704,10.13; 755,6; 769,11; *proêgoumenos*, pre-eminent, 593,16; 769,11; *proêgoumenôs*, pre-eminently, 520,1; 541,14; 733,16; 737,26; 755,6.18; 762,23; 769,13
proïenai, to proceed, go on, 522,23; 529,20; 721,12
prolêpsis, preconception, 529,15; presupposition, 566,24
proödos, procession, 528,17.24
prosekhein, to pay attention to, 576,30; 713,29; 716,32
prosekhes, adjacent, proximate, immediate, closely related, *prosekhôs*, immediately, 532,11.22; 535,24.34; 536,4.11.18-20.35; 537,1-3; 543,30; 549,10; 562,30.31 *passim*
proslêpsis, further assumption, 547,16
protasis, premise, 544,4; 560,28; 705,30; 710,31; 756,11.33.34; *elattôn protasis*, minor premise, 544,2; 559,34; 560,29; 561,34; *meizôn protasis*, major premise, 559,35; 745,20; 756,22
proteros, before, earlier, prior, 519,24; 524,3.5.17.28; 525,27; 526,26; 527,5.9.22; 529,34; 530,20.21.32; 532,34-5
protithenai, to propose, 528,11; 591,11; 695,33; 705,2; 718,19, 747,33; 756,4; 762,29; *protithemenos*, proposed, *to protithemenon*, subject matter, 520,27
prôtôs, primarily, 535,13.15.17.23-8.34.35; 536,1-19; 548,37; 549,2 *passim*
prôtourgos, originally creative, 527,31.35
psukhê, soul, 523,2; 530,12.13; 540,9.30; 546,3; 555,1.5; 564,24; 566,16; 580,27; 592,12.13; 703,28; 708,35; 710,4.8; 713,8; 742,17; 757,7-10; 758,28-32; 759,3-5.16.17.28; 760,6.12-15.18.22.31.34; 761,2.4.7.9; 765,23; 766,14.16.28-31
ptôsis, mode, 566,15

puknôsis, condensation, 588,2; 589,2; 596,11
puknousthai, to become dense, 596,15

rhein, to flow, to run, 578,28; 583,22.33; 716,16; 720,9; 748,16
rhoê, flux, flow, 716,12.14.22.26; 719,26; 738,1.3; 744.19.20.26; 746,4.6
rhuiskesthai, to flow, 722,30
rhusis, flow, 705,8; 722,28; 723,7.10; 724,34

sêmainein, to signify, to mean, 534,14; 552,31; 563,13; 581,28; 584,19; 596,35; 696,5.6.11.17.21; 719,3.6.15.16; 721,32; 725,19; 734,13.14; 735,32; 737,11.28; 739,1; 753,29; *sêmainomenos*, being signified, referred to, *to sêmainomenon*, significance, signification, 522,7; 540,8.10.17.21.29; 541,3.9.21.23; 552,6.7.17; 553,1.5.11; 555,32; 556,21.22; 559,30; 563,22; 569,38; 570,3.9; 591,23 *passim*
sêmeion, indication, sign, 523,13; 549,8.21; point (mathematical), 531,23-30; 590,16; 698,24; 699,15; 712,13.15.31; 713,3; 727,1.11
skhêma, figure, shape, form, 552,26; 573,33; 756,21; 770,26-30; 771,33.34; figure (of a syllogism), 543,2; 547,28; 697,10; 701,6.8; 705,34; 706,3; 713,25.26; 745,17; 756,21
skhêmatismos, arrangement, 742,9
skhesis, position, 521,12; 750,12; respect, 751,26.28; relation, 522,7; 751,31; *en skhesei*, relational, 520,32; in relation, 709,19; *kata skhesin*, s.v. *kata*
skopein, to attend to, examine, consider, 535,2.12; 542,15; 555,22.32.34; 578,25; 597,11; 696,8; 700,14; 703,16; 717,2; 734,22.26
skopos, aim, 578,20; 712,7; 715,9
sôma, body, 519,10-14.19-21.26; 520,1-4.10-13.32; 521,13.20-3; 522,11.20 *passim*; *theion sôma*, divine body, 593,8; 594,23; 758,23
sômatoeidês, bodily, *to somatoeides*, the bodily, 703,20
sphaira, sphere, 531,3; 538,23; 537,4; 580,4; 585,15; 588,7; 589,15.21.30; 590,17.19.25; 591,1.2; 592,1.23; 593,3.8.23; 594,14.19.22.27.29; 597,21; 699,24; 700,19.25.29; 701,10-12; 704,34; 769,20
stadion, stade (a standard of length, *c.* 1/8 of a Roman mile), mile, 709,26; furlong, 735,21; (athletic) track, 712,12; 763,14; mile, *stadiaia*, furlong, 736,20
stasis, being stationary, stationary state, a stop, changelessness, 700,25.26; 726,25; 745,1.2; a stage, 711,11
sterêsis, privation, 519,6; 741,5.6; 742,24-7.34; 743,10; 744,31; 746,2
stoikheion, element, 519,5.17; 520,3; 524,8.12; 529,4.9; 532,2-32.27.30.35.36; 533,4; 534,31 *passim*
sullogismos, syllogism, syllogistic argument, argument, 521,10; 529,29.34; 550,16; 560,30; 580,1; 584,33; 697,9.10; 706,4; 745,17
sullogistikos, logical, 701,5; *sullogistikôs*, syllogistically, 706,8
sumbainein, to happen, 559,8.10; 599,26; 708,17.23; 727,26; 737,11; 753,31; 754,20.33; 769,30; *sumbebêkenai*, to happen to be, to be contingently, 536,15; 543,25.26; 558,10.13.23.31; 559,14; 592,14; 721,14; 728,24; 729,1; 739,1; 742,34; 743,7; 744,4; 746,31; 759,30; 760,31; *to sumbebêkos*, attribute, contingent attribute, 537,14; 543,20.23.24; 552,20; 553,4.9; 554,15; 558,11.18-21.35; 559,15; 560,2.4.5 (see *kata sumbebêkos*)
summethistasthai, to shift with, to change places together, 531,23;

548,11; 576,15; 577,15
sumparatithesthai, to be extended together, correlated, 720,7; 737,19
sumparekteinesthai, to be co-extended, spread alongside, 530,28; 699,4; 741,28
sumperainesthai, to conclude, draw a conclusion, 527,10; 534,29; 535,11; 551,4.7; 560,26; 700,3; 703,20; 709,31; 729,24
sumperasma, conclusion, 529,21; 543,14.31; 544,1; 545,10; 580,1; 745,19
sumperasmatikos, terminal, 752,27
sumphuês, intrinsic, sharing nature, 537,3; 544,19
sumphuesthai, to coalesce, 599,29
sumplêroun, to complete, 519,17; 543,22; to contribute to the whole, 554,15; to compose, 586,13
sunagein, to conclude, draw, draw together, draw to a conclusion, collect, 523,1; 529,5; 542,19; 543,13; 550,8.16; 554,12; 555,26; 559,27.37; 560,6; 565,7.23; 567,9.13 *passim*; *sunagomenos*, following, 706,4; 756,8; *to sunagomenon*, consequence, 537,19
sunagôgê, drawing to a conclusion, 757,25; 759,14; formulation (of an argument), 705,34
sunanairein, to destroy together, annihilate together, abolish together, do away with, 527,34; 528,3; 529,17; 759,2.26; 761,3; 766,33
sunaptein, to tie on to, connect, intr.: touch, lie next to, 543,30; 580,3.5; 582,30; 584,20.24; 593,33; 700,6; 702,29; 708,2; 748,3.20; *sunêmmenos*, connected to, 547,2; *to sunêmmenon*, the conditional, 525,37; 547,16; 574,28; 698,7; the conclusion, the consequent, 531,18; 548,8; 550,21; 556,26; in conditional form, 547,12; a fitting together, 574,23
sundiairein, divide together, 764,9.18; 765,4.9; *sundiaireisthai*, to be distinguished from, 750,29
sundiïstanai, to be coextended, 548,34
sunekheia, continuity, 569,1.25; 711,27.30; 726,16.23; 747,31; 748,12; 749,18.22; 762,4
sunekhein, to make continuous, 726,33; 727,2.3; 748,7.9.10.13.16.29; 749,36
sunekhês, continuous, 519,5; 538,29.35; 568,11.16; 569,15.16,23.26.28; 570,7.14.16.19.26.27.33 *passim*
sunekhizein, to make continuous, produce continuity, 570,6; 573,16; 578,31; 599,29; 726,30
sunêtheia, custom, habit, 722,12; 752,29; experience, 706,7
sungeneia, kinship, 597,19; 599,23.30; 719,25
sungenês, akin, 520,29; 549,16; 597,11.15.23.31.32.35; 598,5.7.14.16.20.24; 599,34-37; 767,30; 768,24
sunnoein, to realise, understand, 546,13; 555,19
sunokhê, collection, 552,5; embrace, 561,6
sunousia, discussion, lecture, 542,10.12; 545,24
suntonos, intense, *suntonos antilêpsis*, continuous application, 522,30; *suntonôs*, intently, 709,13
sustatikos, holding together, constituting, 519,11; 585,23

taktikos, ordinal, 716,19; 718,5
taxis, order, arrangement, 522,7; 541,4-9; 546,13; 580,1; 597,32; 696,11; 702,33; 704,7.17.24; 712,22.24; 714,14.16; 715,4; 717,31; 718,1.4; 762,30-2
tekmêrion, criterion, sign, indication, ground for the truth of something, 566,30; 739,10; 754,25; 758,14
teleios, complete, perfect, 531,3; 598,11; 600,37; 697,31; 703,19; 704,22.28; 737,29

teleiôsis, completion, 742,8
teleiôtikos, perfecting, completing, 704,12; 742,10
teleôs, completely, finally, 529,20; 550,23; 567,14; 571,9
telos, end, 533,3.19-31; 552,13; 556,14; 564,4; 583,33; 590,20; 734,8; 750,25; 755,10.23; 769,32; 770,9
theios, divine, 593,8; 594,23; 758,23
theôrêma, diagram, 526,9; theorem, 532,34
thesis, position, 525,20.28.32.38; 526,3.5.17-21.29.30; 527,12; 529,16; 541,5.7; 568,28; 712,9.30.32; 713,3; 715,5.15-18.21.27.34-6; 716,5.7.11-14.23-5.31 *passim*
tithenai, to place, put forward, posit, suppose, affirm, 520,12; 524,4; 525,37; 527,4; 529,22.25; 532,1.31; 541,2; 545,7; 548,4; 552,8; 553,27; 563,9.14; 600,15.17; 696,24; 701,29; 702,1; 715,26; 763,32; 766,28
tithênê, nurse, 539,11
tonos, intense, 707,13
topikos, providing a place, local, in respect to place, 527,20; 568,21-3.26; 595,21; 715,29; 717,3; 763,24.25; *topikôs*, locally, 568,21.22
topizesthai, to become localised, get a place, 716,10.11
topos, place, 519,20-6; 520,7-13.17.20-4.30.31; 521,1.5.6.11.12.17.27; 522,1-18.27.29 *passim*

Subject Index

References are to the page and line numbers of the *CAG* edition, which appear in the margin of this translation.